An Introduction to FINITE MARKOV PROCESSES

S. R. ADKE
University of Poona, Pune
India

S. M. MANJUNATH
Bangalore University, Bangalore
India

A HALSTED PRESS BOOK

JOHN WILEY & SONS
New York Chichester Brisbane Toronto Singapore

Published in the Western Hemisphere
by Halsted Press, A Division of
John Wiley & Sons, Inc., New York

Library of Congress Cataloging in Publication Data

Printed in India at Gajendra Printing Press, Delhi.

PREFACE

The discrete state-space, continuous parameter Markov process has been used extensively to construct stochastic models in a variety of disciplines like biology, chemistry, electrical engineering, medicine, physics, sociology etc. Most of the books dealing with discrete Markov processes concentrate either on the exposition of their mathematical properties e.g., Chung (1967), Cinlar (1975), Feller (1972), Iosifescu (1980), or provide their introductory properties e.g., Bhat (1972), Karlin and Taylor (1975), Medhi (1982), Parzen (1962). A scientist who employs a finite Markov process to construct a stochastic model has to know, not only the basic assumptions and their consequences, but also the statistical properties and procedures of statistical inference for such processes. This book is intended to meet this requirement.

We discuss the theoretical properties of a finite state-space, continuous parameter Markov process in the first three chapters. Thus chapter 1 provides the basic definitions and establishes the equivalence of various versions of the Markov property. The analytic properties of the transition probabilities, their evaluation and the properties of sample functions are discussed in chapter 2. A detailed discussion of the classification of states of a discrete Markov process is provided in chapter 3 which also discusses the asymptotic behaviour of the transition probabilities. The probability generating function, moment generating function and the first two moments of random variables like duration of stay in a particular state, transition counts for reducible and irreducible Markov processes are discussed in chapter 4. The last and the longest chapter, chapter 5, develops the asymptotic properties of maximum likelihood estimators as well as asymptotic theory of likelihood ratio tests of hypotheses. Finally we provide thirty problems which are constructed on the basis of research papers, some old and some very recent. These problems are based on concepts introduced in two or more chapters and we have therefore listed them at the end of the book.

It is expected that our reader is familiar with basic concepts of probability theory as discussed in Bhat (1981) or in chapters III, IV and VII of Loeve (1968). A good acquaintance with matrix algebra and the contents of chapter V and VI of Rao (1973) is desirable for understanding our chapter 5 on Statistical Inference. This book can be used for a one semester course for advanced level graduate/post graduate students.

Each chapter in the book has been divided into sections which are serially numbered. The definitions, equations, lemmas, theorems, corollaries and examples are all serially numbered in each section. A reference to equation (c) in a section is to the c-th equation of the same section. The equation (b.c) stands for the c-th equation in b-th section of the same chapter and the equation (a.b.c) refers to c-th equation of b-th section of the a-th chapter. The same scheme applies to definitions, lemmas etc.

The book was initiated and almost completed when one of the authors (SMM) was visiting the Department of Statistics, University of Poona as a Teacher Fellow from Bangalore University, Bangalore. We thank Dr A. V. Kharshikar and Dr M. S. Prasad for the discussions we had with them. We also thank Mrs. A. V. Sabane for her careful typing of the manuscript.

University of Poona,
Pune - 7, India
May, 1984.

S. R. ADKE
S. M. MANJUNATH

CONTENTS

CHAPTER 1

DISCRETE MARKOV PROCESSES : DEFINITIONS

1. INTRODUCTION

A large number of natural phenomena which evolve in time have indeterministic components. Such phenomena can be described in terms of an infinite collection of random variables (r.v.s) $X(t)$, $t \in T$, all defined on the same probability space $(\Omega, \mathbb{F}, P)$ and indexed by a parameter t taking values in an infinite index set T; i.e., in terms of the stochastic process $\{X(t), t \in T\}$.

The classical theory of Statistics mainly deals with experiments which are repeatable under identical conditions. The outcomes of such experiments can be modelled in a fairly satisfactory manner by a sequence $\{X_n, n = 1, 2, \ldots\}$ of independent and identically distributed r.v.s. The possibility of dependence between successive r.v.s led Markov (1906) to introduce " chains " of random variables which are now well-known as Markov chains. The sequence of independent and identically distributed r.v.s and the Markov chain are stochastic processes with the set of non-negative integers as the index set. However, the following examples illustrate that it is necessary to consider stochastic processes indexed by a linearly ordered index set like the non-negative half of the real line.

Example 1 : Consider a telephone exchange with a finite number M of channels. Calls arrive at the exchange at random instants of time and a call is connected only if a free channel is available. The availability of a channel depends on the durations of conversations which are also of a random nature. Let $X(t)$ denote the number of busy channels at the instant or the epoch t, $0 \leq t < \infty$. The fluctuations in the random number $X(t)$ of busy channels as t progresses on $[0, \infty)$ are of interest to the design and maintenance engineers. One is obviously dealing with the stochastic process $\{X(t), 0 \leq t < \infty\}$ with $[0, \infty)$ as the index set and $\{0,1,2,..., M\}$ as the set of possible values of $X(t)$, $t \in [0, \infty)$. This example was first discussed by Kolmogorov (1931).

Example 2 : A patient suffering from a disease like cancer can be in a number of different states. The initial state E_1 is the state in which a person is identified as a cancer patient. Depending on the state of his health and the treatment received by him, the patient may move to state E_2 of recovery or to one of the terminal states E_3 or E_4 representing the death of the patient due to cancer or some other cause respectively. A patient may also oscillate between E_1 and E_2 before he is finally claimed by E_3 or E_4 . The state of the patient can be described in terms of the random variable $X(t)$ which equals 1, 2, 3 or 4 according as the state of the patient is E_1, E_2, E_3 or E_4. We are thus dealing with the stochastic process $\{X(t), 0 \leq t < \infty\}$ with $X(t)$ taking values in the set $\{1, 2, 3, 4\}$. This example was discussed by Fix and Neyman (1951).

Example 3 : More recently, Wasserman (1980) has discussed the development of relationships between a group of say M persons. In his most general model, Wasserman defines $X_{ij}(t)$ to be one or zero according as a ' relationship ' exists or does not exist at epoch t between the i-th and j-th individuals of the group, $i \neq j$, $i, j = 1, ..., M$. By convention we take $X_{ii}(t) \equiv 0$, $t \in [0, \infty)$. The entire social network of relationships at any instant t of time can be represented by the M x M matrix $\underline{X}(t) = ((\ X_{ij}(t)\ ))$ of binary elements. The total number of possible binary matrices representing the social network at any instant t is $2^{M(M-1)}$. Wasserman is thus dealing with a matrix-valued stochastic process $\{\underline{X}(t), 0 \leq t < \infty\}$ indexed by the continuous time parameter $t \in [0,\infty)$

As indicated earlier, a stochastic process $\{X(t), t \in T\}$ is an infinite collection of random variables defined on the same probability space $(\Omega, \mathbb{F}, P)$. We shall take the index set T to be either the set Z^+ of non-negative integers or the non-negative half $\mathbb{R}^+ = [0, \infty)$ of the real line $\mathbb{R}$. The parameter t is usually referred to as the time parameter and a point of the index set is called an epoch. The union of range-spaces of the random variables $X(t), t \in T$, is called the state-space S of the stochastic process. If $X(t) = j \in S$, then we say that the stochastic process is in state j at epoch t.

The state-spaces of the stochastic processes in the three examples given above are finite. A process with a finite state-space will be called a finite stochastic process. This book deals with finite stochastic processes which have the Markov property. Loosely speaking, a process has the Markov property if the knowledge of its state at an epoch t is sufficient to determine the probability distribution of $X(u), u > t$; any additional information about $X(s), s < t$, being irrelevant.

In section 2, we define and describe some properties of a Markov chain $\{X_n, n \in Z^+\}$ which are needed in the study of finite Markov processes. The Poisson process and a first definition of a Markov process are introduced in section 3. In section 4, we discuss the different definitions of a Markov process and establish their mutual equivalence and equivalence with the definition introduced in section 3. The strong Markov property is described in section 5. In the last section 6 of this chapter we discuss some properties of the finite dimensional distributions of a finite Markov process.

2. MARKOV CHAINS

Most of the classical theory of Statistics deals with a sequence

$\{X_n, n \in Z^+\}$ of independent and identically distributed (i.i.d.) r.v.s. Suppose the r.v.s $X_n, n \in Z^+$ are non-negative and integer valued. They are said to be identically distributed iff for every $n \in Z^+$,

$$\Pr[X_n = j] = p_j, \qquad j = 0, 1, 2, \ldots,$$

where $p_j \geq 0, j \in Z^+$ and $\sum_{j=0}^{\infty} p_j = 1$. They are independently distributed iff for every $n \geq 2$,

$$\Pr[X_1 = j_1, \ldots, X_n = j_n] = \prod_{r=1}^{n} \Pr[X_r = j_r]$$

for all $j_1, \ldots, j_n \in Z^+$.

A Markov chain constitutes the first weakening of the assumption of i.i.d. nature of the r.v.s. Suppose then that $\{X_n, n \in Z^+\}$ is a sequence of discrete r.v.s taking values in a finite or a countably infinite subset S of the real line $\mathbb{R}$. Such a sequence is said to constitute a Markov chain with state-space S iff for every $n \geq 1, j_0, j_1, \ldots, j_{n+1} \in S$,

$$\Pr[X_{n+1} = j_{n+1} \mid X_0 = j_0, \ldots, X_n = j_n]$$

$$= \Pr[X_{n+1} = j_{n+1} \mid X_n = j_n] \quad , \qquad (1)$$

whenever the conditioning event $[X_0 = j_0, \ldots, X_n = j_n]$ on the left has positive probability. Here we initiate the sequence at n=0 rather than the usual n = 1, because in the study of stochastic processes, it is customary to treat n as a time parameter and to regard X_0 as the r.v. representing state of the initial state of the process. It is for this reason that the distribution of X_0 is usually known as the initial distribution.

Suppose that the initial distribution of a Markov chain $\{X_n, n \in Z^+\}$ is specified by

$$Pr[X_o = j] = a_j, \; j \in S, \; a_j \geq 0, \quad \sum_{j \in S} a_j = 1 ,$$

and let

$$a(n, j, k) = Pr[X_{n+1} = k \mid X_n = j] , \; j, k \in S$$

where $a(n, j, k) \geq 0$ for each $n \in Z^+$ and

$$\sum_{k \in S} a(n, j, k) = 1 , \quad j \in S .$$

A knowledge of $\{a_j, j \in S\}$ and $\{a(n,j,k), n \in Z^+, j, k \in S\}$ enables us to specify the joint distribution of $X_o, X_1, \ldots, X_n$ for each $n \in Z^+$ since

$$\begin{aligned} &Pr[X_o = j_o, \ldots, X_n = j_n] \\ &= Pr[X_o = j_o] \prod_{r=0}^{n-1} Pr[X_{r+1} = j_{r+1} \mid X_o = j_o, \ldots, X_r = j_r] \\ &= a_{j_o} \prod_{r=0}^{n-1} a(r, j_r, j_{r+1}) \qquad (2) \end{aligned}$$

by virtue of the defining equation (1) of the Markov chain.

Example 1 : Ehrenfest Model of Diffusion.

Suppose two urns U_1 and U_2 contain a total of M balls. At each trial, a ball is selected at random out of the M balls, independently of the results of the earlier trials and of the number of balls in U_1 and U_2 at the specific trial. The selected ball is transferred from the urn it is in, to the other urn. Let X_n denote the number of balls in urn U_1 at the end of the n-th trial. Observe that

$$X_{n+1} = X_n + \Delta_{n+1} ,$$

where $\Delta_{n+1} = \pm 1$ according as the selected ball is from U_2 or U_1. It is easy to check that the state-space of the sequence $\{X_n, n \in Z^+\}$ is $S = \{0, 1, ..., M\}$ and that

$$\Pr[X_{n+1} = j_{n+1} \mid X_o = j_o, ..., X_n = j_n]$$

$$= \begin{cases} 1 - j_n/M , & \text{if } j_{n+1} = j_n+1,\ j_n = 0,1,..., M-1, \\ j_n/M , & \text{if } j_{n+1} = j_n-1,\ j_n = 1,2,..., M, \\ 0 , & \text{otherwise}, \end{cases}$$

$$= \Pr[X_{n+1} = j_{n+1} \mid X_n = j_n] .$$

Thus $\{X_n, n \in Z^+\}$ is a Markov chain with state space $S = \{0, 1, ..., M\}$.

In this example observe that for fixed $j, k \in S$,

$$\Pr[X_{n+1} = k \mid X_n = j] = \begin{cases} 1-j/M, & k = j+1,\ j = 0,1,..., M-1, \\ j/M , & k = j-1,\ j = 1,2,..., M, \\ 0 , & \text{otherwise} ; \end{cases}$$

for all $n \geq 0$. In other words, the conditional probability of the event $[X_{n+1} = k]$, given the event $[X_n = j]$, does not depend on $n \varepsilon Z^+$. This observation leads to the following

<u>Definition 1</u> : A Markov chain $\{X_n, n \varepsilon Z^+\}$ is said to have <u>stationary transition probabilities</u> or to be <u>time homogeneous</u> if the conditional probability $Pr[X_{n+1} = k \mid X_n = j]$ does not depend on $n \varepsilon Z^+$ for all $j, k \varepsilon S$.

Hereinafter, unless otherwise specified, a Markov chain will be assumed to have stationary transition probabilities. Let

$$p_{jk} = Pr[X_{n+1} = k \mid X_n = j], \quad j, \quad k \varepsilon S,$$

which is the probability of a transition from state j to state k in one step and is therefore called a <u>one-step transition probability</u>. These probabilities obviously satisfy the following conditions :

$$p_{jk} \geq 0, \; j, k \varepsilon S, \quad \sum_{k \varepsilon S} p_{jk} = 1, \quad j \varepsilon S.$$

If the state-space S is finite, we may take $S = \{1, 2, \ldots, M\}$ without loss of generality and arrange p_{jk} in the form of a square matrix $P = ((p_{jk}))$ of order M, whose (j, k) - element in the j-th row and k-th column is p_{jk}. The matrix P of one-step transition probabilities is thus a stochastic matrix in the sense of the following

<u>Definition 2</u> : A square matrix A of order M is a <u>stochastic matrix</u> if all its elements a_{jk} are non-negative and for each

$$j = 1, \ldots, M, \quad \sum_{k=1}^{M} a_{jk} = 1.$$

We denote the n-step transition probability

$Pr[X_{m+n} = k \mid X_m = j]$, $j, k \varepsilon S, n \varepsilon Z^+$, by $p_{jk}^{(n)}$, with the obvious convention that

$$p_{jk}^{(o)} = \delta_{jk} = \begin{cases} 1, & \text{if } j = k , \\ 0, & \text{otherwise} , \end{cases}$$

and $p_{jk}^{(1)} = p_{jk}$. Observe that

$$p_{jk}^{(n+1)} = \Pr[X_{m+n+1} = k \mid X_m = j]$$

$$= \sum_{r \in S} \Pr[X_{m+n} = r \mid X_m = j] . \Pr[X_{m+n+1} = k \mid X_{m+n} = r, X_m = j] \quad (3)$$

$$= \sum_{r \in S} p_{jr}^{(n)} \; p_{rk} , \quad (4)$$

where (3) is a consequence of the theorem of total probabilities and (4) is a consequence of the Markovian nature of the sequence $\{X_n, n \in Z^+\}$. The equation (4) provides us with a recursive way of calculating the n-step transition probabilities.

One can use the above argument also to establish that

$$p_{jk}^{(m+n)} = \sum_{r \in S} p_{jr}^{(m)} \; p_{rk}^{(n)} = \sum_{r \in S} p_{jr}^{(n)} \; p_{rk}^{(m)} \quad (5)$$

for all $m, n \in Z^+$, $j, k \in S$. These equations play an important role in the study of Markov chains and are called the Chapman - Kolmogorov equations.

If $P^{(n)} = ((p_{jk}^{(n)}))$ denotes the matrix of n-step transition probabilities, then the Chapman-Kolmogorov equations (5) become

$$P^{(n+m)} = P^{(n)} P^{(m)} = P^{(m)} P^{(n)} \tag{6}$$

in matrix notation. It is obvious that $P^{(n)}$ is also a stochastic matrix and that in fact it is the n-th power P^n of the matrix P of one-step transition probabilities. In this connection one may observe that by virtue of equation (2), the initial distribution and the one-step transition probabilities of a Markov chain determine the joint distribution of $X_o, \ldots, X_n$ for all $n \in Z^+$ and hence as a consequence, the joint distribution of any finite subset of the sequence $\{X_n, n \in Z^+\}$.

An important aspect of the study of a Markov chain is the classification of its states. We proceed to describe the classification of the states and refer to Chung (1967) for details.

A state j <u>leads</u> to a state k, $j \to k$, if there exists an $n \in Z^+$ such that $P_{jk}^{(n)} > 0$. Thus, by definition, every state leads to itself as $p_{jj}^{(o)} = 1$. Two states j and k <u>communicate</u> if they lead to each other. A proper subset C of the state-space is a <u>closed</u> set if $j \in C$, $k \notin C$, implies that $p_{jk}^{(n)} = 0$ for all $n \in Z^+$. The state space is closed by definition. A closed set C is <u>minimal closed</u> if no proper subset of C is closed. A Markov chain is <u>irreducible</u> iff its state-space is minimal closed. If for a state j, $p_{jj} = 1$, then j is said to be an <u>absorbing state.</u> It is easy to verify that for an absorbing state j, $p_{jk}^{(n)} \equiv 0$ for all $k \neq j$ and all $n \in Z^+$.

A state j is an <u>essential</u> state if it communicates with every state it leads to; i.e. if $p_{jk}^{(n)} > 0$ implies the existence of an $m \in Z^+$ such that $p_{kj}^{(m)} > 0$. A state which is not essential is an inessential state.

Example 2 : Suppose $\{X_n, n \in Z^+\}$ is a Markov chain with one-step transition probability matrix

$$P = \begin{pmatrix} 1 & 0 & 0 & 0 \\ 0 & 1/2 & 1/2 & 0 \\ 0 & 1/3 & 2/3 & 0 \\ 1/4 & 1/4 & 1/4 & 1/4 \end{pmatrix} .$$

It is easy to check that in this chain, state 4 leads to states 1, 2 and 3 but that state 1 does not lead to state 4. Hence state 4 is inessential. States 2 and 3 communicate with each other and are essential states forming a minimal closed set. The state 1 is an absorbing state.

Observe that in Example 1, if we take M = 1, we get the trivial Markov chain on {0, 1} with $p_{01} = p_{10} = 1$, so that

$$p_{00}^{(2n)} = 1 , \quad p_{00}^{(2n+1)} = 0 , \quad n \in Z^+ .$$

Thus the greatest common divisor (g.c.d.) of the set $\{n \mid p_{00}^{(n)} > 0 , n \geq 1\}$ is 2 and therefore state zero may be said to be periodic with period 2. More generally, a state j is periodic with period d, if the g.c.d. of the set $\{n \mid p_{jj}^{(n)} > 0 , n \geq 1\}$ is $d > 1$. A state j for which d = 1 is called an aperiodic state.

Further classification of states of a Markov chain is based on the probability of its return to its original state. Define

$$f_{jk}^{(1)} = p_{jk}$$

and

$$f_{jk}^{(n)} = \Pr[X_n = k,\ X_r \neq k,\ 0 < r < n \mid X_0 = j]$$

$$= \Pr[X_{m+n} = k,\ X_{m+r} \neq k,\ 0 < r < n \mid X_m = j]\ ,\ n \geq 2.$$

Observe that $f_{jj}^{(n)}$ is the probability of first return of the chain to state j at epoch n and since these events are mutually exclusive,

$$f_{jj} = \sum_{n=1}^{\infty} f_{jj}^{(n)}$$

is the probability of at least one return of the chain to the state j. We, therefore, say that a state j is persistent if $f_{jj} = 1$ and transient if $f_{jj} < 1$. A persistent state is non-null or null according as its mean recurrence time $\mu_j = \Sigma\ n\ f_{jj}^{(n)} < \infty$ or $+\infty$. An aperiodic, persistent and non-null state is called an ergodic state.

It is possible to prove that for an aperiodic state j, $\lim_{n\to\infty} p_{jj}^{(n)} = 0$ whenever j is persistent null or transient. If j is ergodic and $i \to j$, then $\lim_{n\to\infty} p_{ij}^{(n)} = 1/\mu_j > 0$. One immediate consequence of this fact is that a finite Markov chain, has to have at least one persistent, non-null state and that none of its states can be null.

Let j be a fixed state and let $C_{(j)}$ be the set of states which communicate with j. Such a set is called a communicating class or simply a class of states. A property α of a state is said to be a solidarity property or a class property if the possession of the property α by one state in the class implies that all states in that class have the property α. It can be shown that being essential, inessential, periodic, aperiodic, transient, persistent null, persistent non-null are all solidarity properties.

3. POISSON PROCESS AND DISCRETE MARKOV PROCESS

Consider random events such as disintegration of nuclear particles, arrival of telephone calls, chromosome breakages under harmful irradiations, accidents at a specified intersection in a busy city, occurence of defects along the length of an electrical cable etc. Let $N(t)$ denote the number of occurrences of such an event in the interval $[0, t)$. If the successive occurrences of the event take place under similar conditions, it is reasonable to assume that the stochastic process $\{N(t), 0 \leq t < \infty\}$ is a process with stationary and independent increments according to the following

Definition 1 : A stochastic process $\{X(t), 0 \leq t < \infty\}$ is a process with stationary and independent increments if for all $n \geq 1$, $0 = t_0 < t_1 < \ldots < t_n$, the increments $X(t_1) - X(0), X(t_2) - X(t_1), \ldots, X(t_n) - X(t_{n-1})$ are such that

(i) their joint distribution is the same as that of $X(t_1+h) - X(h), X(t_2+h) - X(t_1+h), \ldots, X(t_n+h) - X(t_{n-1}+h)$ for all $h > 0$, and

(ii) they are independently distributed in the sense that

$$Pr[X(t_1) - X(0) \leq x_1, \ldots, X(t_n) - X(t_{n-1}) \leq x_n]$$

$$= \prod_{j=1}^{n} Pr[X(t_j) - X(t_{j-1}) \leq x_j]$$

for all real $x_1, \ldots, x_n$.

It is obvious that the state-space of $\{N(t), 0 \leq t < \infty \}$ is the set of non-negative integers and since $N(t)$ counts the number of occurrences of the event in [0, t), we must have $N(s) \leq N(t)$ whenever $s \leq t$. Such a process is, for obvious reasons called a <u>counting process</u>.

<u>Definition 2</u> : A counting process $\{N(t), 0 \leq t < \infty\}$ with stationary and independent increments such that

$$\text{(i)} \quad Pr[N(0) = 0] = 1$$

$$\text{(ii)} \quad Pr[N(t) = r] = \exp(-\lambda t)(\lambda t)^r/r! \ , \ r = 0,1,2,\ldots,\lambda \in (0,\infty),$$

is called a <u>Poisson process</u> with <u>rate</u> λ.

We bring out an important property of the Poisson process in the following

<u>Lemma 1</u> : Let $\{N(t), 0 \leq t < \infty\}$ be a Poisson process according to definition 2. Let $0 = t_o < t_1 < t_2 < \ldots < t_n < t_{n+1} < \infty$ be points in IR^+ and $0 \leq r_1 \leq r_2 \leq \ldots \leq r_n \leq r_{n+1}$ be points in Z^+. Then

$$Pr[N(t_{n+1}) = r_{n+1} \mid N(t_1) = r_1, \ldots, N(t_n) = r_n]$$

$$= Pr[N(t_{n+1}) = r_{n+1} \mid N(t_n) = r_n] . \qquad (1)$$

<u>Proof</u> : In view of the almost sure non-decreasing nature of the Poisson process, the probabilities in (1) are meaningful only under the specified order restrictions on $r_1, \ldots, r_{n+1}$. We have

$$Pr[N(t_1) = r_1, \ldots, N(t_{n+1}) = r_{n+1}]$$

$$= Pr[N(t_1)-N(0)=r_1, N(t_2)-N(t_1)=r_2-r_1,\ldots),N(t_{n+1})-N(t_n)=r_{n+1}-r_n]$$

$$= \prod_{j=1}^{n-1} \Pr[N(t_j) - N(t_{j-1}) = r_j - r_{j-1}] \qquad (2)$$

$$= \frac{\prod_{j=1}^{n+1} [\exp[-\lambda(t_j - t_{j-1})]] \{\lambda(t_j - t_{j-1})\}^{(r_j - r_{j-1})}}{(r_j - r_{j-1})!} , \qquad (3)$$

where (2) is a consequence of the independence of the increments and (3) is a consequence of their stationarity and definition 2. A similar expression can be obtained for

$\Pr[N(t_1) = r_1, \ldots, N(t_n) = r_n]$ so that

$$\Pr[N(t_{n+1}) = r_{n+1} \mid N(t_1) = r_1, \ldots, N(t_n) = r_n]$$

$$= \frac{\exp\{-\lambda(t_{n+1} - t_n)\}\{\lambda(t_{n+1} - t_n)\}^{r_{n+1} - r_n}}{(r_{n+1} - r_n)!}$$

$$= \Pr[N(t_{n+1}) = r_{n+1} \mid N(t_n) = r_n] ,$$

which completes the proof of the lemma.

The equation (1) can be looked upon as a generalization to the continuous parameter case of the equation (2.1) defining a Markov chain. We may thus say that a Poisson process is a Markov process. This enables us to introduce

Definition 3 : A continuous parameter stochastic process $\{X(t), 0 \leq t < \infty\}$ defined on a complete probability space

$(\Omega, \mathbb{F}, P)$ is said to be a <u>discrete Markov process</u> if its state-space S is either a finite or a countably infinite subset of $\mathbb{R}$ and if for every $n \geq 2$, $0 \leq t_1 < \ldots < t_n < t_{n+1} < \infty$, $i_1, \ldots, i_n, i_{n+1} \in S$,

$$\Pr[X(t_{n+1}) = i_{n+1} | X(t_1) = i_1, \ldots, X(t_n) = i_n]$$

$$= \Pr[X(t_{n+1}) = i_{n+1} | X(t_n) = i_n] \qquad (4)$$

whenever the conditioning event $[X(t_1) = i_1, \ldots, X(t_n) = i_n]$ on the left of (4) has positive probability.

<u>Remark</u> : A discrete Markov process whose state-space is a finite subset of $\mathbb{R}$ is called a <u>finite</u> Markov process.

We now describe an easy method of constructing a discrete Markov process from a discrete Markov chain. Let $\{Y_n, n \in Z^+\}$ be a Markov chain with one-step transition probabilities p_{jk}, $j, k \in S$, and let $\{N(t), 0 \leq t < \infty\}$ be an independent Poisson process. Define

$$X(t) = Y_{N(t)} , \quad 0 \leq t < \infty . \qquad (5)$$

<u>Lemma 2</u> : The process $\{X(t), 0 \leq t < \infty\}$, where $X(t)$ is defined by equation (5), is a discrete Markov process.

<u>Proof</u> : It is obvious that the state spaces of $\{X_n, n \in Z^+\}$ and $\{X(t), 0 \leq t < \infty\}$ processes are the same. Let

$0 \leq t_1 < \ldots < t_{n+1} < \infty$ and let $j_1, \ldots, j_n, j_{n+1} \in S$.

Consider

$$\Pr[X(t_1) = j_1, \ldots, X(t_n) = j_n, X(t_{n+1}) = j_{n+1}]$$

$$= \sum_{r_1=0}^{\infty} \cdots \sum_{r_{n+1}=0}^{\infty} \Pr[Y_{r_1} = j_1, Y_{r_1+r_2} = j_2, \ldots, Y_{r_1+\ldots+r_{n+1}} = j_{n+1}]$$

$$\times \Pr[N(t_1) = r_1, N(t_2) = r_1+r_2, \ldots, N(t_{n+1}) = r_1+\ldots+r_{n+1}] ,$$

which is a consequence of the independence of the $\{Y_n\}$ - and $\{N(t)\}$- processes. Using the Markov nature of the $\{X_n\}$- process and equation (3), one has

$$\Pr[X(t_1) = j_1, \ldots, X(t_{n+1}) = j_{n+1}]$$

$$= \sum_{r_o=0}^{\infty} \Pr[Y_{r_1} = j_1] \exp(-\lambda t_1) (\lambda t_1)^{r_1}/r_1 !$$

$$\prod_{\nu=2}^{n+1} [\sum_{r_\nu=0}^{\infty} p_{j_{\nu-1} j_\nu}^{(r_\nu)} \exp\{-\lambda(t_\nu - t_{\nu-1})\}\{\lambda(t_\nu - t_{\nu-1})\}^{r_\nu}/r_\nu!] .$$

It now readily follows, after an easy computation, that

$$\Pr[X(t_{n+1}) = j_{n+1} \mid X(t_1) = j_1, \ldots, X(t_n) = j_n]$$

$$= \sum_{r=0}^{\infty} p_{j_n j_{n+1}}^{(r)} \exp[-\lambda(t_{n+1} - t_n)] \{\lambda(t_{n+1} - t_n)\}^r/r!$$

$$= \Pr[X(t_{n+1}) = j_{n+1} \mid X(t_n) = j_n] . \tag{6}$$

Thus $\{X(t), 0 \leq t < \infty\}$ is a discrete Markov process.

Suppose that $\{Y_n, n \varepsilon Z^+\}$ is a Markov chain with state-space S and $\{N(t), 0 \leq t < \infty\}$ is an independent Poisson process. The discrete Markov process $\{X(t), 0 \leq t < \infty\}$, where $X(t)$ is defined by equation (5), is called the derived Markov process, derived from the Markov chain $\{Y_n, n \varepsilon Z^+\}$ and the Poisson process is called the deriving process. [cf. Stam (1965)] . Sometimes the $\{X(t)\}$-process is said to be the $\{X_n\}$-chain governed by a Poisson clock or a random clock. [cf. Lamperti (1977)]. We now proceed to analyse the definition 3 of a discrete Markov process.

In analogy with the definition of the initial distribution of a Markov chain, we shall call the distribution of $X(0)$ as the <u>initial distribution</u> of the discrete Markov process. If

$$\Pr[X(0) = j] = p_j(0) ,$$

where $p_j(0) \geq 0$, $j \varepsilon S$ and $\sum_{j \varepsilon S} p_j(0) = 1$, is the initial distribution, then for any $t > 0$,

$$\Pr[X(t) = j] = \Sigma \Pr[X(0) = i] \Pr[X(t) = j | X(0) = i].$$

Moreover, if $t_1 \ldots t_n$ are any n ordered epochs, the joint distribution of $X(t_1), \ldots, X(t_n)$ is specified by

$$\Pr[X(t_1) = j_1, \ldots, X(t_n) = j_n]$$

$$= \Pr[X(t_1) = j_1] \prod_{\nu=1}^{n-1} \Pr[X(t_{\nu+1}) = j_{\nu+1} | X(t_\nu) = j_\nu], j_1, \ldots, j_n \varepsilon S.$$

Thus one can obtain the joint distribution of any finite number of r.v.s of the discrete Markov process in terms of its initial distribution and the probabilities

$$p_{jk}(s, t) = Pr[X(t) = k \mid X(s) = j], j, k \in S, 0 \leq s < t < \infty .$$

The probability $p_{jk}(s,t)$ is said to be the transition probability of a transition from state j at epoch s to state k at epoch $t > s$ and the function $p_{jk}(s, t)$, $j, k \in S$, $0 \leq s < t < \infty$, defined on $S \times S \times T$, where $T = \{(s, t) \mid 0 \leq s < t < \infty\}$ is called the transition probability function (t.p.f.) of the discrete Markov process.

It is easy to see from equation (6) that for the derived Markov process

$$p_{jk}(s, t) = \sum_{r=0}^{\infty} p_{jk}^{(r)} \exp[-\lambda(t-s)] . [\lambda(t-s)]^r/r! ,$$

so that the t.p.f. of the derived Markov process depends on s and t only through their difference t-s. Thus, for every $h > 0$,

$$p_{jk}(s, t) = Pr[X(t+h) = k \mid X(s+h) = j] . \qquad (7)$$

This leads to the following

Definition 4 : A discrete Markov process is said to have stationary transition probabilities or to be time homogeneous iff its t.p.f. is such that for all $j, k \in S$ and $s > 0$,

$$Pr[X(t) = k \mid X(0) = j] = Pr[X(s+t) = k \mid X(s) = j] . \qquad (8)$$

Hereinafter, we shall restrict ourselves to those discrete Markov processes which have stationary transition probabilities and write $p_{jk}(t)$ for the transition probability defined by equation (8).

Convention : In analogy with the convention for Markov chains, define

$$p_{jk}(0) = \delta_{jk} = \begin{cases} 1, & \text{if } j = k , \\ 0, & \text{otherwise} . \end{cases}$$

We illustrate the different ideas introduced in this section with the following examples.

Example 1 : Poisson process. The Poisson process is a Markov process whose t.p.f. is specified by

$$p_{jk}(t) = \begin{cases} \exp(-\lambda t)\ (\lambda t)^{j-k}/(j-k)! & \text{if } k \geq j,\ j = 0,1,2,\ldots, \\ 0 , & \text{otherwise.} \end{cases}$$

It is easy to verify that the Poisson process itself can be looked upon as a Markov process derived from the Markov chain on Z^+ whose one step transition probabilities are

$$p_{jk} = \begin{cases} 1, & \text{if } k = j+1,\ j \in Z^+ \\ 0, & \text{otherwise} . \end{cases}$$

Example 2 : Let $\{Y_n, n \in Z^+\}$ be a finite Markov chain with state-space $\{1, 2, \ldots, M\}$ and transition probability matrix P. The t.p.f. of the derived Markov process is specified by

$$p_{jk}(t) = \sum_{n=0}^{\infty} p_{jk}^{(n)} \exp(-\lambda t)\ (\lambda t)^n/n\ !$$

and therefore the main problem in obtaining $p_{jk}(t)$ for a specific Markov chain is that of evaluating the powers of P. A simple solution is available when P has distinct characteristic roots i.e., when the M roots $\alpha_1, \ldots, \alpha_M$ of the determinantial equation

$$|P - \alpha I_M| = 0$$

are such that $\alpha_i \neq \alpha_j$ for $i \neq j$, $i, j = 1, \ldots, M$, I_M being the unit matrix of order M. Let ξ_r and η_r be $M \times 1$ column vectors denoting the right and left characteristic vectors of P corresponding to the characteristic root α_r :

$$P \xi_r = \alpha_r \xi_r \, , \quad \eta_r^T P = \alpha_r \eta_r^T \, , \; r = 1, \ldots, M,$$

where A^T denotes the transpose of the matrix A. Let $U = (\xi_1, \ldots, \xi_M)$ and $V = (\eta_1, \ldots, \eta_M)$ be $M \times M$ matrices whose columns are the right and left characteristic vectors of P. The above equations can be written in matrix notation as

$$PU = U\Lambda \, , \quad V^T P = \Lambda V^T \, ,$$

where $\Lambda = \text{diag}(\alpha_1, \ldots, \alpha_M)$ is the diagonal matrix with $\alpha_1, \ldots, \alpha_M$ along the main diagonal. The distinctness of $\alpha_1, \ldots, \alpha_M$ implies that the right vectors $\xi_1, \ldots, \xi_M$ as well as the left vectors $\eta_1, \ldots, \eta_M$ are linearly independent. Thus U and V are non-singular matrices. Moreover,

$$\alpha_i \eta_i^T \xi_j = \eta_i^T P \xi_j = \alpha_j \eta_i^T \xi_j \, , \qquad i \neq j \, ,$$

and therefore $(\alpha_i - \alpha_j) \eta_i^T \xi_j = 0$ implies that $\eta_i^T \xi_j = 0$ for $i \neq j$. The matrix $V^T U$ is, therefore, a diagonal, non-singular matrix. Hence $\eta_i^T \xi_i \neq 0$, $i = 1, \ldots, M$ and we may assume, after a suitable rescaling if necessary, that $\eta_i^T \xi_i = 1$, $i = 1, \ldots, M$.

It follows that

$$P = U\Lambda V^T = \sum_{r=1}^{M} \alpha_r \xi_r \eta_r^T \tag{9}$$

from which one easily has

$$P^n = \sum_{r=1}^{M} \alpha_r^n \xi_r \eta_r^T , \quad n \in Z^+ ,$$

using the fact that $\xi_i^T \eta_j = 0$ for $i \neq j$. The equation (9) is called the spectral resolution of P.

Let $a_{jk}^{(r)}$ denote the (j,k)-element of the M x M matrix $\xi_r \eta_r^T$, $r = 1, \ldots, M$. Then

$$p_{jk}^{(n)} = \sum_{r=1}^{M} \alpha_r^n a_{jk}^{(r)}$$

from which it follows that

$$p_{jk}(t) = \sum_{r=1}^{M} a_{jk}^{(r)} \exp\{- \lambda(1 - \alpha_r)t\} . \tag{10}$$

In obtaining (9) we have not used the fact that P is a stochastic matrix. One can show that unity is always a characteristic root of P and that all the characteristic roots of P lie within the unit circle centred at the origin of the complex plane.

Let $\alpha_1 = 1$ and suppose that $|\alpha_j| < 1$, $j = 2,3,\ldots,M$. Then

$$p_{jk}^{(n)} = a_{jk}^{(1)} + \Sigma \alpha_r^n a_{jk}^{(r)}$$

$$\to a_{jk}^{(1)} \quad \text{as} \quad n \to \infty .$$

Since $\alpha_1 = 1$, we may take $\xi_1 = (1,\ldots,1)^T$ and η_1 such that $\eta_1^T \xi_1 = 1$. Thus $a_{jk}^{(1)}$ which is the (j,k)-element of $\xi_1 \eta_1^T$ does not depend on j and we may write $a_{jk}^{(1)} = \pi_k$, and under the stated conditions, $\lim_{n\to\infty} p_{jk}^{(n)} = \pi_k$ not depending on j. In particular it also follows that

$$\lim_{t\to\infty} p_{jk}(t) = \pi_k, \quad j, k = 1, \ldots, M .$$

We shall have more to say about the limit behaviour of $p_{jk}(t)$ as $t \to \infty$ in section 3.5.

<u>Example 3</u> : Suppose that a mechanism can be either in a working state E_0 or in the state E_1 of undergoing repair. Let $X(t)$ be zero or one according as the mechanism is in E_0 or E_1 at epoch $t \in \mathbb{R}^+$. Assume that the process $\{X(t), t \in \mathbb{R}^+\}$ is a Markov process such that

$$p_{01}(h) = p_{01}(t, t+h) = \lambda h + o(h) ,$$
$$p_{10}(h) = p_{10}(t, t+h) = \mu h + o(h) , \qquad (11)$$

where λ and μ are positive constants and $o(h)$ denotes a function of $h \in [0, \infty)$ with the property that $o(h)/h \to 0$ as $h \downarrow 0$. The process being a Markov process, we must have

$$p_{00}(t+h) = \Pr[X(t+h) = 0 \mid X(0) = 0] .$$

$$= p_{00}(t)\ p_{00}(h) + p_{01}(t)\ p_{10}(h)$$

$$= [1 - \lambda h + o(h)]\ p_{00}(t) + [\mu h + o(h)]\ p_{01}(t)$$

from which it follows that

$$\{p_{00}(t+h) - p_{00}(t)\}/h = -\lambda p_{00}(t) + \mu p_{01}(t) + o(h)/h . \quad (12)$$

As $h \downarrow 0$, the right side of (12) converges and therefore so must the left side of (12) which is, in fact, the right derivative of $p_{00}(t)$. If, therefore, one assumes that $p_{00}(t)$ is differentiable on $(0, \infty)$, one gets the following differential equation :

$$dp_{00}(t)/dt = -\lambda p_{00}(t) + \mu p_{01}(t) \quad (13)$$

Similar arguments and assumptions lead to

$$dp_{01}(t)/dt = -\mu p_{01}(t) + \lambda p_{00}(t) ,$$

$$dp_{10}(t)/dt = -\lambda p_{10}(t) + \mu p_{11}(t) ,$$

$$dp_{11}(t)/dt = -\mu p_{11}(t) + \lambda p_{10}(t) ,$$

which are to be solved subject to the natural conditions

$$p_{00}(0) = p_{11}(0) = 1, \quad p_{01}(0) = p_{10}(0) = 0 .$$

The equation (13) can be rewritten as

$$dp_{00}(t)/dt = -(\lambda+\mu)\, p_{00}(t) + \mu$$

which is easily solved to obtain

$$p_{00}(t) = \mu/(\lambda+\mu) + \lambda \exp[-(\lambda+\mu)t]/(\lambda+\mu) .$$

It is now easy to verify, by symmetry, that

$$p_{11}(t) = \lambda/(\lambda+\mu) + \mu \exp[-(\lambda+\mu)t]/(\lambda+\mu) .$$

Suppose that the initial distribution of the process is

$$Pr[X(0) = 0] = \alpha , \quad Pr[X(0) = 1] = (1-\alpha) , \quad \alpha \in [0,1].$$

Then

$$Pr[X(t) = 0] = \alpha\, p_{00}(t) + (1-\alpha)\, p_{10}(t)$$

$$\to \mu/(\lambda+\mu)$$

and therefore $P[X(t) = 1] \to \lambda/(\lambda+\mu)$, as $t \to \infty$. Thus $X(t)$ converges in distribution, as $t \to \infty$, to a limit random variable whose distribution does not depend on the initial distribution.

We shall find in section 2.2 that when dealing with a discrete Markov process, it is not necessary to make assumptions of the type specified by equation (11), and that they are consequence of much simpler and natural assumptions.

4. ALTERNATIVE DEFINITIONS OF A MARKOV PROCESS AND THEIR EQUIVALENCE

The definition 3.3 of a discrete Markov process is not its only definition. The Markov property of a continuous parameter process can be expressed in a number of different ways. We now describe the different ways in which the Markov property can be defined and establish their equivalence. Although in this book we are concerned with a finite state-space, our difficulties do not increase if we assume that the state-space S of the process is any arbitrary Borel subset of $\mathbb{R}$. In what follows $\{X(t),\ t \in \mathbb{R}^+\}$ is a continuous parameter, real stochastic process defined on a probability space $(\Omega, \mathbb{F}, P)$ which is assumed to be complete.

We did not need any sophisticated definition of the conditional probability in introducing definition 3.3 of a discrete Markov process as it is expressed in terms of the conditional probability of an event given another event of positive probability. However, in our subsequent development we need the concept and properties of the conditional expectation of a r.v. given a σ-field.

We say that a r.v. X on $(\Omega, \mathbb{F}, P)$ is integrable if it has a finite expectation $E\ X$ or equivalently if $E|X| < \infty$. Let X be an integrable r.v. and $\mathbb{B}$ be a sub σ-field of $\mathbb{F}$. The <u>conditional expectation</u> $E(X|\mathbb{B})$ of X, given the σ-field $\mathbb{B}$, is any $\mathbb{B}$- measurable r.v. such that for every $B \in \mathbb{B}$

$$\int_B E(X|\mathbb{B})\, dP = \int_B X\, dP \quad . \tag{1}$$

The <u>conditional probability</u> $P(A|\mathbb{B})$ of an event A, given the

σ-field $\mathbb{B}$ is, by definition the conditional expectation $E(I(A) \mid \mathbb{B})$ of the indicator function $I(A)$ of the event A. The conditional expectation $E(X|Y_t, t \in T)$ of a r.v. X, given the r.v.s $Y(t)$, $t \in T$, defined on $(\Omega, \mathbb{F}, P)$, T being an arbitrary non - empty index set, is the conditional expectation of X given the σ-field $\sigma\{Y(t), t \in T\}$ induced by the r.v.s Y_t, $t \in T$. In fact $\sigma\{Y(t), t \in T\}$ is by definition the minimal σ-field over $\bigcup_{t \in T} \sigma\{Y(t)\}$.

The conditional expectation $E(X|\mathbb{B})$ is almost surely uniquely defined in the sense that if Z_1 and Z_2 are any two $\mathbb{B}$-measurable r.v.s satisfying (1), then $Pr[Z_1 = Z_2] = 1$ or equivalently there exists a P-null set N such that for every $\omega \in N^c$, $Z_1(\omega) = Z_2(\omega)$. We shall denote this fact by writing $Z_1 = Z_2$ a.s. [P]. It is presumed that the reader is familiar with the elementary properties of conditional expectations as developed in chapter 12 of Bhat (1981). Apart from the usual linearity properties of conditional expectations we shall need their following smoothing properties.

(i) If X is a $\mathbb{B}$-measurable r.v. and Y is a r.v. such that XY and Y are integrable, then

$$E(XY \mid \mathbb{B}) = X\, E(Y \mid \mathbb{B}) \quad \text{a.s. [P]} . \tag{2}$$

(ii) If $\mathbb{B}_1 \subset \mathbb{B}_2$ and $\mathbb{B}_1$, $\mathbb{B}_2$ are sub σ-fields of $\mathbb{F}$, then

$$E(E(X \mid \mathbb{B}_2) \mid \mathbb{B}_1) = E(X \mid \mathbb{B}_1) = E(E(X \mid \mathbb{B}_1) \mid \mathbb{B}_2) \tag{3}$$

with probability one.

A simple modification of definition 3.3 is provided by

<u>Definition 1</u> : The stochastic process $\{X(t), t \in \mathbb{R}^+\}$ is a Markov process if for any u and t, $0 \leq t < u < \infty$, and any Borel set E,

$$P\{X(u) \in E \mid X(s), s \leq t\} = P\{X(u) \in E \mid X(t)\} \tag{4}$$

almost surely [P] .

It is customary to designate the σ-field $\sigma\{X(s), s < t\}$ as the past, $\sigma\{X(t)\}$ as the present and $\sigma\{(X(u), u > t\}$ as the future. The definition 1 is then interpreted to assert that the future depends on the past only through the present. An apparanently more general definition of a Markov process is

<u>Definition 2</u> : The stochastic process $\{X(t), t \in \mathbb{R}^+\}$ is a Markov process if for any event $A \in \sigma\{X(u), u > t\}$,

$$P\{ A \mid X(s), s \leq t \} = P\{ A \mid X(t) \} \tag{5}$$

almost surely

One would expect that if the future depends on the past only through the present, then the past should also depend on the future only through the present. Moreover, a more symmetric way of defining a Markov process would be to assert that the past and the future are independent given the present. These heuristic statements are made precise in the following two definitions.

<u>Definition 3</u> : The stochastic process $\{X(t), t \in \mathbb{R}^+\}$ is a Markov process if for any $t > 0$ and any event $B \in \sigma\{X(s), s < t\}$,

$$P\{B \mid X(s), s \geq t\} = P\{ B \mid X(t)\} \tag{6}$$

almost surely [P] .

Definition 4 : If for any $t > 0$, the σ-fields $\sigma\{X(s), s > t\}$ and $\sigma\{X(s), s < t\}$ are conditionally independent, i.e., if for any $A \in \sigma\{X(s), s > t\}$ and $B \in \sigma\{X(s), s < t\}$

$$P\{A \cap B \mid X(t)\} = P\{A \mid X(t)\} P\{B \mid X(t)\} \qquad (7)$$

with P-probability 1, then the process $\{X(t), t \in \mathbb{R}\}$ is a Markov process.

Remark : Let ξ_1, ξ_2 and ξ_3 be integrable r.v.s on $(\Omega, \mathbb{F}, P)$ such that ξ_1 is $\sigma\{X(u)$ -measurable for some $u > t$, ξ_2 is $\sigma\{X(s), s > t\}$ -measurable and ξ_3 is $\sigma\{X(s), s < t\}$ -measurable. Then equations (4), (5) and (6) are respectively equivalent to the following equations which also hold almost surely [P] .

$$E(\xi_1 \mid X(s), s \leq t) = E(\xi_1 \mid X(t)) , \qquad (4')$$

$$E(\xi_2 \mid X(s), s \leq t) = E(\xi_2 \mid X(t)) , \qquad (5')$$

$$E(\xi_3 \mid X(s), s \geq t) = E(\xi_3 \mid X(t)) . \qquad (6')$$

These assertions being valid for appropriate indicator functions, they can be extended to simple, non-negative r.v.s. One can then use the conditional form of the Lebesgue monotone convergence theorem to extend the result to non-negative r.v.s. The final extention to real, integrable r.v.s is immediate.

The equations (4) - (7) and (3.4) provide us with five equivalent versions of what is known as the Markov property of the stochastic process $\{X(t), t \in \mathbb{R}^+\}$. Our first major exercise is to establish that the above four definitions of a Markov process are equivalent. We proceed to do this.

Lemma 1 : Let $\mathbb{F}_1$ and $\mathbb{F}_2$ be two σ-fields of subsets of Ω. Let

$$\mathbb{D} = \{A_1 \cap A_2 \mid A_i \in \mathbb{F}_i \ , \ i = 1, 2\}$$

and

$$\mathbb{D}^* = \{\bigcup_{i=1}^{n} H_i \mid H_i \in D, \ i = 1, \ldots, n, \ n \in Z^+\}.$$

Then $\mathbb{D}^*$ is the minimal field over $\mathbb{D}$. Moreover, the minimal σ-field $\sigma\{\mathbb{D}\}$ over $\mathbb{D}$ and the minimal σ-field $\sigma\{\mathbb{F}_1 \cup \mathbb{F}_2\}$ over $\mathbb{F}_1 \cup \mathbb{F}_2$ coincide.

Proof : It is easy to see that $\mathbb{D} \subset \mathbb{D}^*$ and that $\mathbb{D}^*$ is closed under finite unions. Let $B = \bigcup_{i=1}^{n} H_i \in \mathbb{D}^*$, with $H_i = A_{i1} \cap A_{i2}$, $A_{ij} \in \mathbb{F}_j$, $j = 1, 2$, $i = 1, \ldots, n$. Then $B^c = \bigcap_{i=1}^{n} \{A_{i1}^c \cup A_{i2}^c\}$.

If $n = 1$, then

$$B^c = A_{11}^c \cup A_{12}^c = (A_{i1}^c \cap \Omega) \cup (A_{i2}^c \cap \Omega) \in \mathbb{D}^* .$$

One can now use induction to claim that $\mathbb{D}^*$ is closed under complementation. Thus $\mathbb{D}^*$ is a field over $\mathbb{D}$, and since it consists of sets which are finite unions of subsets in $\mathbb{D}$, it is a sub-field of every other field containing $\mathbb{D}$ i.e., $\mathbb{D}^*$ is the minimal field over $\mathbb{D}$.

In order to show that $\sigma\{\mathbb{D}\} = \sigma\{\mathbb{F}_1 \cup \mathbb{F}_2\}$, observe that $\mathbb{D} \supset \mathbb{F}_1 \cup \mathbb{F}_2$ and therefore $\sigma\{\mathbb{D}\} \supset \sigma\{\mathbb{F}_1 \cup \mathbb{F}_2\}$. On the other hand, by construction, $\mathbb{D} \subset \mathbb{D}^* \subset \sigma\{\mathbb{F}_1 \cup \mathbb{F}_2\}$ so that

$\sigma\{\mathbb{D}\} \subset \sigma\{\mathbb{F}_1 \cup \mathbb{F}_2\}$. This establishes the fact that $\sigma\{\mathbb{D}\} = \sigma\{\mathbb{F}_1 \cup \mathbb{F}_2\}$.

If X is an integrable r.v., then

$$\varphi_X(A) = \int_A X \, dP = \int_\Omega XI(A) \, dP$$

is defined for each $A \in \mathbb{F}$ and is called the indefinite integral of X on $\mathbb{F}$. If X is non-negative and integrable, then $\varphi_X(.)$ is a finite measure on $\mathbb{F}$.

Lemma 2 : Let X and Y be non-negative and integrable r.v.s on $(\Omega, \mathbb{F}, P)$ and let $\mathbb{F}_1$ and $\mathbb{F}_2$ be two sub σ-fields of $\mathbb{F}$. If the indefinite integrals $\varphi_X(.)$ and $\varphi_Y(.)$ of X and Y respectively, coincide on $\mathbb{D}$, then they coincide on $\sigma\{\mathbb{F}_1 \cup \mathbb{F}_2\}$.

Proof : Observe that φ_X and φ_Y being σ-additive on $\mathbb{F}$, their equality on $\mathbb{D}$ implies their equality on $\mathbb{D}^*$. But by the Caratheodory extention theorem [cf. Loève (1968), p. 87-90] and by Lemma 1 above it follows that $\varphi_X = \varphi_Y$ on $\sigma\{\mathbb{D}^*\} = \sigma\{\mathbb{F}_1 \cup \mathbb{F}_2\}$.

Theorem 1 : The definitions 2, 3 and 4 of a Markov process are equivalent.

Proof : We first establish that definitions 2 and 4 are equivalent. Let $A \in \sigma\{X(s), s > t\}$ and $B \in \sigma\{X(s), 0 \le s < t\}$. We have to prove that

$$E\{I(A) I(B) \mid X(t)\} = E\{I(A) \mid X(t)\} E\{I(B) \mid X(t)\}$$

almost surely [P] iff

$$E\{I(A) \mid X(s),\ s \leq t\} = E\{I(A) \mid X(t)\} \text{ a.s. } [P] . \qquad (8)$$

Since $\sigma\{X(t)\} \subset \sigma\{X(s),\ s \leq t\}$ and $I(B)$ is a $\sigma\{X(s),\ s \leq t\}$ -measurable r.v., the smoothing properties (2) and (3) of the conditional expectations imply that

$$E\{I(A)\ I(B) \mid X(t)\} = E[I(B)\ E\{I(A) \mid X(s),\ s \leq t\} \mid X(t)] \text{ a.s. } [P]$$

and that

$$E\{I(A) \mid X(t)\}\ E\{I(B) \mid X(t)\}$$

$$= E[I(B)\ E\{I(A) \mid X(t)\} \mid X(t)] \text{ a.s. } [P] .$$

Thus it is enough to prove that (8) holds iff

$$E[I(B)\ E\{I(A) \mid X(s) \leq t\} \mid X(t)]$$

$$= E[I(B)\ E\{I(A) \mid X(t)\} \mid X(t)] \qquad (9)$$

with P-probability one for all $A \in \sigma\{X(s),\ s > t\}$ and $B \in \sigma\{X(s),\ s < t\}$.

Suppose (8) holds. Multiply both sides of (8) by $I(B)$ and take conditional expectation, given $X(t)$, to obtain (9).

Conversely, suppose (9) holds. Then by the definition of conditional expectation, for every $H \in \sigma\{X(t)\}$

$$\int_H I(B)\ E\{I(A) \mid X(s),\ s \leq t\}\, dP$$

$$= \int_H I(B)\ E\{I(A) \mid X(t)\}\, dP \quad .$$

Equivalently

$$\int_{B\cap H} E\{I(A) \mid X(s),\ s \leq t\}dP = \int_{B\cap H} E\{I(A) \mid X(t)\}\ dP$$

holds for every $H \in \sigma\{X(t)\}$ and $B \in \sigma\{X(s),\ s \leq t\}$. An application of lemma 2, with

$$X = E\ \{I(A) \mid X(s),\ s \leq t\}\ ,\quad Y = E\ \{I(A) \mid X(t)\}$$

$$\mathbb{F}_1 = \sigma\{X(s),\ s \leq t\}\ ,\quad \mathbb{F}_2 = \sigma\{X(t)\}$$

implies that the indefinite integrals of the non-negative integrable r.v.s $E\{I(A) \mid X(s),\ s \leq t\}$ and $E\{I(A) \mid X(t)\}$ coincide on the σ-field $\sigma\{X(s),\ s \leq t\}$, with respect to which both of them are measurable. Hence (8) holds almost surely [P]. This completes the proof of equivalence of definition 2 and 4 of the Markov process. The proof of the equivalence of definitions 3 and 4 is similar and is therefore omitted. The theorem is established.

In order to establish the equivalence of definitions 1 and 2 of a Markov process, we need the following

<u>Lemma 3</u> : Let T be a non-empty subset of $\mathbb{R}^+$ and let $\sigma\{X(s),\ s \in T\}$ be the σ-field generated by the r.v.s $X(s),\ s\in T$. Let

$$\mathbb{D} = \{B_{t_1} \cap \cdots \cap B_{t_n} \mid B_{t_j} \in \sigma\{X(t_j)\},\ j = 1,\ldots,n\ ,$$
$$(t_1,\ \ldots,\ t_n) \subset T\ ,\quad n \in Z^+\}$$

denote the class of all finite intersections of the events in

$\bigcup_{s \in T} \sigma\{X(s)\}$. Let

$$\mathbb{D}^* = \{\bigcup_{j=1}^{n} H_j \mid H_j \in \mathbb{D} \ , \ j = 1,\dots, n, \ n \in Z^+ \}$$

denote the class of all finite unions of events in $\mathbb{D}$. Then $\mathbb{D}^*$ is the minimal field over $\mathbb{D}$. Moreover, the minimal σ-field over $\mathbb{D}$ and $\sigma\{X(s), s \in T\}$ coincide .

Proof : The proof of this lemma is similar to the proof of lemma 1 except for notational complications and is therefore omitted.

Theorem 2 : The definitions 1 and 2 of a Markov process are equivalent.

Proof : If $u > t$ and E is any Borel set, the event $[X(u) \in E] \in \sigma\{X(s), s > t\}$ and therefore definition 2 implies definition 1.

Suppose now that definition 1 holds, $u_2 > u_1 > t$ and let E_1, E_2 be Borel sets. Then

$$P[X(u_1) \in E_1, X(u_2) \in E_2 \mid X(s), \ s \le t \}$$

$$= E[E\{I(X(u_1) \in E_1) \ I(X(u_2) \in E_2) \mid X(s) \le u_1\} \mid X(s), \ s \le t]$$

$$= E[I(X(u_1) \in E_1) \ E\{I(X(u_2) \in E_2) \mid X(s), \ s \le u_1\} \mid X(s), \ s \le t]$$

by virtue of the facts that (i) $\sigma\{X(s), s \le u_1\} \supset \sigma\{X(s), s \le t\}$ (ii) $I[X(u_1) \in E_1]$ is $\sigma\{X(s), s \le u_1\}$-measurable and the smoothing properties. Using definition 1 for $E\{I(X(u_2) \in E_2) \mid X(s), s \le u_1\}$

we obtain

$$P[X(u_1) \varepsilon E_1, \ X(u_2) \varepsilon E_2 \mid X(s), \ s \leq t]$$

$$= E[I(X(u_1) \varepsilon E_1) \ P[X(u_2) \varepsilon E_2 \mid X(u_1)] \mid X(s), \ s \leq t]$$

$$= E\{I(X(u_1) \varepsilon E_1) \ P[X(u_2) \varepsilon E_2 \mid X(u_1)] \mid X(t)\}$$

by virtue of (4). Repeating the above argument with $P[X(u_1) \varepsilon E_1, \ X(u_2) \varepsilon E_2 \mid X(t)]$, one arrives at the conclusion that for $t < u_1 < u_2$, Borel sets E_1, E_2, definition (1) implies that

$$P[X(u_1) \varepsilon E_1, \ X(u_2) \varepsilon E_2 \mid X(s), \ s \leq t]$$

$$= P[X(u_1) \varepsilon E_1, X(u_2) \varepsilon E_2 \mid X(t)] \text{ a.s. } [P] .$$

A simple induction argument implies that for any n Borel sets $E_1, \ldots, E_n$, and real numbers $0 < t < u_1 < \ldots < u_n$,

$$P\{\bigcap_{j=1}^{n} [X(u_j) \varepsilon E_j \mid X(s), \ s \leq t\}$$

$$P\{\bigcap_{j=1}^{n} [X(u_j) \varepsilon E_j] \mid X(t)\} \quad \text{a.s.} \quad [P] . \tag{10}$$

Thus if definition (1) holds, then (5) is true for all sets of the type

$$\bigcap_{j=1}^{n} [X(t_j) \varepsilon E_j] .$$

We must extend this conclusion to any set $E \in \sigma\{X(s), s > t\}$. It is enough to show that

$$\int_A P[E \mid X(t)]\, dP = P[A \cap E] \tag{11}$$

for all $A \in \sigma\{X(s), s \leq t\}$.

For fixed $A \in \sigma\{X(s), s \leq t\}$, let $\mathbb{D}_A$ be the class of sets $E \in \sigma\{X(s), s > t\}$ for which (11) holds. It is not difficult to verify that $\mathbb{D}_A$ is a σ-field. But by (10), $\mathbb{D}_A$ contains all sets of the form $\bigcap_{j=1}^{n} [X(u_j) \in B_j]$, where B_j are Borel sets. In particular $\mathbb{D}_A$ must contain the σ-field generated by the class

$$\{ \bigcap_{j=1}^{n} [X(u_j) \in B_j] \mid t < u_1 < \ldots < u_n , j = 1,\ldots,n, n \in Z^+ \}$$

which by lemma 3, is in fact the σ-field $\sigma\{X(s), s > t\}$. Since this assertion is true for each $A \in \sigma\{X(s), s \leq t\}$, we conclude that (11) is true for all $A \in \sigma\{X(s), s \leq t\}$ and $E \in \sigma\{X(s), s > t\}$. This completes the proof of the theorem.

We now connect definition 3.3 with the four definitions of a Markov process discussed in this section. Observe that the event $[X(t_{n+1}) = j_{n+1}] \in \sigma\{X(t_{n+1})\}$ and therefore one can introduce the following

<u>Definition 5</u> : The process $\{X(t), t \in \mathbb{R}^+\}$ is a Markov process if for every $n \geq 1$, real numbers $0 \leq t_1 < \ldots < t_n < t < u < \infty$, and any Borel set E,

$$P[X(u) \in E \mid X(t_1), \ldots, X(t_n), X(t)] = P[X(u) \in E \mid X(t)] \tag{12}$$

almost surely [P].

When the state-space is discrete, definition 5 immediately implies definition 3.3. Conversely, for a discrete-state space, the event $[X(u) \in E]$ is either empty or is an event of the type $\sum_{k \in E \cap S} [X(u) = k]$. Thus definition 3.3 implies definition 5. It is thus enough to establish that definition 5 is equivalent to definitions 1, 2, 3 and 4. This we accomplish in the following

Theorem 3 : The definitions 1-5 of a Markov process are equivalent.

Proof : In view of theorems 1 and 2, it is enough to establish that definitions 1 and 5 are equivalent.

Observe that $\sigma\{X(t_1), \ldots, X(t_n), X(t)\} \subset \sigma\{X(s), s \leq t\}$. Hence if definition 1 holds then equation (12) and therefore definition 5 hold by a simple application of the smoothing property of the conditional expectation.

Conversely, suppose that definition 5 holds. Then for all

$H \in \sigma\{X(t_1), \ldots, X(t_n), X(t)\}$

$$\int_H P[X(u) \in E \mid X(t)]dP = P[H \cap [X(u) \in E]] . \qquad (13)$$

It is thus enough to prove that (13) holds for all $A \in \sigma\{X(s), s \leq t\}$. Let $A_j \in \sigma\{X(t_j)\}$, $j = 1, \ldots, n+1$ with $t_{n+1} = t$. Then it is obvious that (11) holds for $\bigcap_{j=1}^{n+1} A_j$ as well as for finite unions of events of the type $\bigcap_{j=1}^{n+1} A_j$. This is evidently also true for all finite subsets $\{t_1, \ldots, t_n, t\} \subset [0, t]$.

However, by using lemma 3, one can show that the minimal σ-field over the class of all such events is the σ-field $\sigma\{X(s), s \leq t\}$. The fact that definition 5 implies definition 1 now follows by an application of the Caratheodory extention theorem as in lemma 2.

In view of the equivalence of the different definitions of a Markov process, we shall refer to the properties defined by equations (3.4), (4), (5), (6), (7) and (12) simply as the Markov property. In the next section we introduce the concept of the so-called strong Markov property.

5. STRONG MARKOV PROPERTY

Let $\{X(t), t \in \mathbb{R}^+\}$ be a stochastic process on a complete probability space $(\Omega, \mathbb{F}, P)$. Let $\mathbb{F}_t$ denote the σ-field generated by the r.v.s $X(u)$, $u \leq t$. The definition 4.1 of a Markov process is the same as the assertion that $\{X(t), t \in \mathbb{R}^+\}$ has the Markov property if for any Borel set E and $t > 0$,

$$P[X(t_o + t) \in E \mid \mathbb{F}_{t_o}] = P[X(t_o + t) \in E \mid X(t_o)] \text{ a.s.}[P]. \tag{1}$$

Thus, loosely speaking, a process has the Markov property if given $X(t_o) = x$, the process begins afresh for $t > t_o$, with x as its initial value. The question we ask is whether this property holds if t_o is replaced by a random epoch. Such a question becomes meaningful in the study of the properties of the sample functions of a stochastic process.

Definition 1 : Let $\{X(t), t \in \mathbb{R}^+\}$ be a real stochastic process on a complete probability space $(\Omega, \mathbb{F}, P)$. Let $\omega \in \Omega$ be fixed. Then the function $X(., \omega)$ with domain $\mathbb{R}^+$ is called a sample function or a sample path of the process $\{X(t), t \in \mathbb{R}^+\}$.

In the study of the sample functions of a stochastic process random times occur in an inevitable manner. One might be interested in the epoch of first discontinuity of a sample function or the epoch of first attainment of a specific value by the process. One would expect that the Markov property should hold if t_0 in (1) is replaced by a random epoch since the Markov property does hold for each value of the random epoch. We proceed to make these ideas precise.

Definition 2 : Let $\{\mathbb{H}_t , t \in \mathbb{R}^+\}$ be an increasing family of sub-σ-fields of the σ-field $\mathbb{F}$ i.e., let $\mathbb{H}_s \subset \mathbb{H}_t \subset \mathbb{F}$ for every $0 \leq s < t < \infty$. The stochastic process $\{X(t), t \in \mathbb{R}^+\}$ is said to be adapted to the family $\{\mathbb{H}_t, t \in \mathbb{R}^+\}$ if $X(t)$ is $\mathbb{H}_t$-measurable for every $t \in \mathbb{R}^+$

By definition every process is adapted to the family $\{\mathbb{F}_t , t \in \mathbb{R}^+\}$. In fact it is adapted to any bigger family $\{\mathbb{H}_t, t \in \mathbb{R}^+\}$, if $\mathbb{H}_t \supset \mathbb{F}_t$, $t \in \mathbb{R}^+$. We obtain a stronger version of the Markov property in the following

Definition 3 : The real stochastic process $\{X(t), t \in \mathbb{R}^+\}$ is said to have the Markov property with respect to the increasing family $\{\mathbb{H}_t , t \in \mathbb{R}^+\}$ of sub σ-fields of $\mathbb{F}$ if

(i) $\{X(t), t \in \mathbb{R}^+ \}$ is adapted to $\{\mathbb{H}_t, t \in \mathbb{R}^+\}$,

(ii) for every $t \in \mathbb{R}^+$, $B \in \sigma\{X(u), u > t\}$,

$$P[B \mid \mathbb{H}_t] = P[B \mid X(t)] \text{ a.s.} \qquad (2)$$

This property defined above is stronger than the usual Markov property because a process obeying the definition 3 has the usual Markov property, but the converse may not be true.

In what follows we shall be mainly concerned with an

increasing family $\{\mathbb{H}_t, t \in \mathbb{R}^+\}$ of sub σ-fields which is right continuous in the sense that $\bigcap_{s>t} \mathbb{H}_s = \mathbb{H}_t$ for every $t \in \mathbb{R}^+$. This assumption removes some minor technical difficulties in later developments.

Definition 4 : A positive r.v. τ (which may be extended real - valued) on a probability space $(\Omega, \mathbb{F}, P)$ is said to be a stopping time with respect to an increasing, right continuous family $\{\mathbb{H}_t, t \in \mathbb{R}^+\}$ of sub σ-fields of $\mathbb{F}$ if for every $x > 0$, the event $[\tau \leq x] \in \mathbb{H}_x$.

Associated with every stopping time τ, is the σ-field

$$\mathbb{F}_\tau = \{A \mid A \in \mathbb{F}_\infty , A \cap [\tau \leq x] \in \mathbb{F}_x \text{ for every } x > 0 \}$$

where $\mathbb{F}_\infty$ is the smallest σ-field containing $\bigcup_{t \in \mathbb{R}^+} \mathbb{F}_t$. Observe further that $X(\tau)$ is a r.v. on $(\Omega, \mathbb{F}, P)$ defined by the relation

$$X(\tau)(\omega) = X(\tau(\omega), \omega) \quad . \qquad (3)$$

We are now in a position to define the strong Markov property.

Definition 5 : Let $\{X(t), t \in \mathbb{R}^+\}$ be a stochastic process with Markov property with respect to the right continuous increasing family $\{\mathbb{H}_t, t \in \mathbb{R}^+\}$ of sub σ-fields. It is said to have the strong Markov property with respect to $\{\mathbb{H}_t, t \in T\}$ if for any Borel set E

$$P[X(\tau + u) \in E \mid \mathbb{H}_\tau] = P[X(\tau + u) \in E \mid X(\tau)]$$

for all positive r.v.s τ which are stopping times with respect to $\{\mathbb{H}_t, t \in \mathbb{R}^+\}$.

We refer the interested reader to Chung (1967), p. 172-182 for a proof of the fact that a finite Markov process has the strong Markov property with respect to the family $\{\mathbb{F}_t, t \in \mathbb{R}^+\}$.

6. FINITE DIMENSIONAL DISTRIBUTIONS

Let $\{Y(t), t \in \mathbb{R}^+\}$ be a real stochastic process with continuous time parameter. Most of the statistical properties of such a process can be obtained if we know the joint distribution of $Y(t_1), \ldots, Y(t_n)$ for all $n = 1, 2, \ldots$ and all $t_1, \ldots, t_n \in \mathbb{R}^+$. If

$$F_n(y_1,\ldots,y_n; t_1, \ldots, t_n) = \Pr[\bigcap_{j=1}^{n} [Y(t_j) \leq y_j]], \tag{1}$$

$y_1, \ldots, y_n \in \mathbb{R}$, denotes the joint distribution function of $Y(t_1), \ldots, Y(t_n)$, then the collection

$$\{F_n(\cdot, \ldots, \cdot ; t_1,\ldots,t_n) \mid 0 \leq t_1 < \ldots < t_n < \infty , n=1,2,\ldots\}$$

of all such distribution functions is called the <u>family of finite dimensional distributions</u> associated with the stochastic process $\{Y(t), t \in \mathbb{R}^+\}$. If all the r.v.s $Y(t)$ are discrete, we may specify the joint probability mass function

$$\Pr[Y(t_1) = i_1, \ldots, Y(t_n) = i_n] \tag{2}$$

for all $i_1, \ldots, i_n$ in the state-space of the process and for all $t_1, \ldots, t_n \in \mathbb{R}^+$ instead of specifying F_n as in (1) and (2) determine each other in case of discrete r.v.s.

<u>Theorem 1</u> : The initial distribution

$$p_i(0) = \Pr[X(0) = i], \quad i \in S \tag{3}$$

and the transition probabilities

$$p_{ij}(t) = \Pr[X(u+t) = j \mid X(u) = i] , \ i, j \in S, \ u, t \in \mathbb{R}^+ \qquad (4)$$

of a discrete Markov process $\{X(t), t \in \mathbb{R}^+\}$ with state-space S determine its family of finite dimensional distributions.

Proof : Since the r.v.s $X(t)$, $t \in \mathbb{R}^+$, are all discrete, it is enough to demonstrate that (3) and (4) determine

$$\Pr[X(t_1) = i_1, \ldots, X(t_n) = i_n]$$

for all $i_1, \ldots, i_n \in S$, all $t_1, \ldots, t_n \in \mathbb{R}^+$, $0 \leq t_1 < \ldots < t_n$ and all $n \geq 1$. Let $n = 1$, replace i_1 and t_1 by i and t respectively and observe that

$$\Pr[X(t) = i] = \sum_{s \in S} \Pr[X(0) = s] \Pr[X(t) = i \mid X(0) = s]$$

$$= \sum_{s \in S} p_s(0) \; p_{si}(t) \qquad (5)$$

Now let $n > 1$ and observe that

$$\Pr[X(t_1) = i_1, \ldots, X(t_n) = i_n]$$

$$= \Pr[X(t_1) = i_1] \prod_{\nu=2}^{n} \Pr[X(t_\nu) = i_\nu \mid X(t_1)=i_1, \ldots, X(t_{\nu-1})=i_{\nu-1}]$$

$$= \Pr[X(t_1) = i_1] \prod_{\nu=2}^{n} \Pr[X(t_\nu) = i_\nu \mid X(t_{\nu-1}) = i_{\nu-1}] \qquad (6)$$

$$= \{\sum p_s(0) \; p_{si_1}(t_1)\} \prod_{\nu=2}^{n} p_{i_{\nu-1} i_\nu}(t_\nu - t_{\nu-1}) , \qquad (7)$$

where (6) is a consequence of the Markov property. The equations

(5) and (7) complete the proof of the theorem.

We now proceed to obtain an analogue of the Chapman-Kolmogorov equations (2.5) for a discrete Markov process. They play a very important role in the study of finite Markov processes.

Lemma 1 : The transition probabilities of a discrete Markov process satisfy the Chapman - Kolmogorov equations

$$p_{jk}(t+u) = \sum_{r\varepsilon S} p_{jr}(u)\, p_{rk}(t) = \sum_{r\varepsilon S} p_{jr}(t)\, p_{rk}(u) \tag{8}$$

for all $j, k \varepsilon S$, $u, t \varepsilon (0, \infty)$.

Proof : We have

$$P_{jk}(t+u) = Pr[X(t+u+h) = k \mid X(h) = j]$$

$$= \sum_{r\varepsilon S} Pr[X(u+h) = r|X(h) = j]\, Pr[X(t+u+h) = k|X(u+h) = r, X(h) = j] \tag{9}$$

$$= \sum_{r\varepsilon S} Pr[X(u+h) = r|X(h) = j]\, Pr[X(t+u+h) = k|X(u+h) = r] \tag{10}$$

$$= \sum_{r\varepsilon S} p_{jr}(u)\, p_{rk}(t) \tag{11}$$

where (9) is a consequence of the theorem of total probabilities, and (10) is a consequence of the Markov property. The other relation in (8) is established in a similar manner.

Remark : If S is the finite set $\{1, 2, \ldots, M\}$ and $P(t) = ((p_{jk}(t)))$ denotes the matrix of transition probabilities, then equation (8) become

$$P(t+u) = P(t)\,P(u) = P(u)\,P(t) \qquad (12)$$

in matrix notation. We shall refer to $P(t)$ as the transition probability matrix. Observe that, $P(t)$ is obviously a stochastic matrix for each $t \in (0, \infty)$.

It is interesting to note that the Chapman-Kolmogorov equations (8) or (12), although a consequence of the Markov property, do not characterise it. One can construct a non-Markov process $\{Y(t), t \in \mathbb{R}^+\}$ whose transition probabilities satisfy the Chapman-Kolmogorov equations. The following lemma, which is the converse of lemma 3.2, is needed for the purpose of illustrating the preceding remark.

<u>Lemma 2</u> : Let $\{X_n, n \in Z^+\}$ be a sequence of discrete r.v.s and let $\{N(t), t \in \mathbb{R}^+\}$ be an independent Poisson process with parameter λ. If

$$Y(t) = X_{N(t)}, \quad t \in \mathbb{R}^+,$$

is a discrete Markov process, then $\{X_n, n \in Z^+\}$ is a Markov chain.

<u>Proof</u> : Let $T_1, T_1 + T_2, \ldots, T_1 + T_2 + \ldots + T_n, \ldots$ denote the successive random epochs of the jumps of the Poisson process $\{N(t), t \in \mathbb{R}^+\}$, i.e., let

$$T_1 + \ldots + T_n = \inf \{t \mid N(t) = n\}, \; n = 1,2,\ldots \qquad (13)$$

It is well-known [cf. Parzen (1962), p. 135] that the sequence $\{T_n, n \geq 1\}$ of r.v.s is a sequence of i.i.d. r.v.s with common density function.

$$f(x) = \lambda \exp(-\lambda x), \quad 0 \leq x < \infty .$$

By construction,

$$X_n = Y(T_1 + \dots + T_n) , \quad n = 1, 2, \dots$$

so that for $j_o, \dots, j_n \in S$, the state-space of the sequence $\{X_n, n \in Z^+\}$,

$$Pr[X_o = j_o, X_1 = j_1, \dots, X_n = j_n]$$

$$= Pr[Y(0) = j_o, Y(T_1) = j_1, \dots, Y(T_1+\dots+T_n) = j_n]$$

$$= \int_0^\infty \dots \int_0^\infty Pr\,[Y(0) = j_o, Y(t_1) = j_1, \dots, Y(t_n) = j_n]$$

$$\lambda^n \exp\{-\lambda \sum_{j=1}^n t_j\}\, dt_1 \dots dt_n$$

$$= Pr[Y(0) = j_o] \prod_{r=1}^n \int_0^\infty p_{j_{r-1}j_r}(t_r)\, \lambda \exp(-\lambda t_r)\, dt_r$$

where $p_{jk}(t)$ denotes the transition probability associated with the $Y(t)$ - process. It is now easy to check that

$$Pr[X_n = j_n \,|\, X_o = j_o, \dots, X_n = j_n] = Pr[X_n = j_n \,|\, X_{n-1} = j_{n-1}] ,$$

which concludes the proof of the lemma.

This lemma enables us to discuss the following example due to Feller (1959).

Example 1 : Let $M \geq 3$ be a fixed integer. Let C_1 be the set of M! permutations of 1, 2, ..., M and let $C_2 = \{(k,\dots,k) \,|\, k=1,\dots,M\}$ be the set containing M elements, each of which is the M-fold repetition $(k, \ldots, k)$ of an integer k, $1 \leq k \leq M$. Define $\Omega = C_1 \cup C_2$ and let $\mathbb{F}$ be the class of all the subsets of Ω. Assign probability $(1 - M^{-1})/M!$ to each point in C_1 and probability M^{-2} to each point of C_2. Let P denote the probability measure so obtained on the measurable space $(\Omega, \mathbb{F})$.

Define $X_n(\omega)$ to be the n-th co-ordinate of $\omega \in \Omega$, $n = 1, 2, \ldots, M$. Thus $X_1, \ldots, X_M$ are M r.v.s on the probability space $(\Omega, \mathbb{F}, P)$. In order to obtain a sequence of r.v.s, we let $X_{rM+1}, \ldots, X_{(r+1)M}$, $r = 1, 2, \ldots,$ to be independent copies of $X_1, \ldots, X_M$.

Define X_o to be a r.v. independent of X_n, $n \geq 1$, such that $\Pr[X_o = j] = M^{-1}$, $j = 1, \ldots, M$. Thus we have a sequence $\{X_n, n \in Z^+\}$ of r.v.s defined on an appropriate probability space.

It is readily verified that

$$\Pr[X_m = j] = M^{-1}, \quad \Pr[X_m = j, X_n = k] = M^{-2}, \quad m \neq n,$$

so that the n-step transition probability

$$p_{jk}^{(n)} = \Pr[X_{m+n} = k \mid X_m = j] = M^{-1}, \qquad j, k = 1, \ldots, M, \; n \geq 1 .$$

The Chapman-Kolmogorov equations 2.5 are easily verified. Moreover, if $j \neq k$,

$$\Pr[X_3 = k \mid X_2 = j, X_1 = j] = 0 \neq \Pr[X_3 = k \mid X_2 = j] = M^{-1}$$

clearly establishing that $\{X_n, n \in Z^+\}$ is not a Markov chain.

Let $\{N(t), t \in \mathbb{R}^+\}$ be a Poisson process with parameter

$\lambda = 1$ and let $\{N(t), t \in \mathbb{R}^+\}$ and $\{X_n, n \in Z^+\}$ be independent processes. It is easy to see that if $Y(t) = X_{N(t)}$, $t \in \mathbb{R}^+$, then

$$\Pr[Y(t) = j] = M^{-1}, \qquad j = 1, \ldots, M$$

and

$$\Pr[Y(u) = j, Y(t+u) = k] = \begin{cases} M^{-1}e^{-t} + M^{-2}(1-e^{-t}), & j = k \\ M^{-2}(1-e^{-t}), & j \neq k, \end{cases}$$

so that

$$p_{jk}(t) = \Pr[Y(t+u) = k \mid Y(u) = j]$$

$$= \begin{cases} e^{-t} + M^{-1}(1 - e^{-t}), & j = k, \\ M^{-1}(1 - e^{-t}), & j \neq k, \; j, k = 1, \ldots, M. \end{cases}$$

One can now easily observe that the Chapman-Kolmogorov equations (8) hold in spite of the fact that the $Y(t)$ - process is not Markovian by virtue of lemma 2.

We have already seen in theorem 1 that the initial distribution and the transition probabilities of a discrete Markov process completely determine its family of finite dimensional distributions. Suppose now that S is a finite or a countably infinite subset of $\mathbb{R}$. Let $\{a_j, j \in S\}$ be a set of non-negative numbers such that $\sum_{j \in S} a_j = 1$. Let $a_{jk}(t)$ be defined for all j, $k \in S$ and $t \in (0, \infty)$, such that $a_{jk}(t) \geq 0$ $\sum_{k \in S} a_{jk}(t) = 1$ and that they satisfy the Chapman-Kolmogorov equations (8). The following theorem asserts that there exists a discrete Markov process with initial distribution $\{a_j, j \in S\}$ and transition probabilities $a_{jk}(t)$.

Theorem 2 : Let S, a_j, $a_{jk}(t)$, $j, k \in S$, $t \in (0, \infty)$ be as defined in the preceding paragraph. Then there exist a probability space $(\Omega, \mathbb{F}, P)$ and a discrete Markov process $\{X(t), t \in \mathbb{R}^+\}$ defined on $(\Omega, \mathbb{F}, P)$ such that for all $j, k \in S$, $t, u \in (0, \infty)$,

$$P[X(0) = j] = a_j, \quad P[X(u+t) = k \mid X(u) = j] = a_{jk}(t) \quad .$$

Proof : Define $p_j(0) = a_j$, and let

$$p_j(t) = \sum_{i \in S} a_i \, a_{ij}(t) \; , \quad t \in (0, \infty), \; j \in S \; .$$

Let S^n denote the n-fold Cartesian product of S with itself, $(j_1, \ldots, j_n) \in S^n$, $0 \leq t_1 < \ldots < t_n < \infty$, $n \geq 1$. Define

$$\Psi_1(j, t) = p_j(t) \quad ,$$

$$\Psi_n(j_1,\ldots,j_n;t_1,\ldots,t_n) = p_{j_1}(t_1) \prod_{r=2}^{n} a_{j_{r-1}j_r}(t_r - t_{r-1}), \; n \geq 2, \tag{14}$$

which are easily seen to be one and n-dimensional probability mass functions respectively. It is not difficult to show that Ψ_n's lead to a family of finite dimensional distribution functions F_n satisfying the following conditions of consistency and symmetry.

(i) Condition of consistency : For all $n > 1$, $t_1, \ldots, t_n \in \mathbb{R}^+$ $0 \leq t_1 < \ldots < t_n < \infty$

$$\lim_{x_n \to \infty} F_n(x_1, \ldots, x_n; t_1, \ldots, t_n) = F_{n-1}(x_1, \ldots, x_{n-1}; t_1, \ldots, t_{n-1}).$$

(ii) Condition of symmetry : Let $(t_{i_1}, \ldots, t_{i_n})$ be a permutation of $(t_1, \ldots, t_n)$ and let $(x_{i_1}, \ldots, x_{i_n})^n$ be the corresponding permutation of $(x_1, \ldots, x_n)$. Then

$$F_n(x_{i_1}, \ldots, x_{i_n}, t_{i_1}, \ldots, t_{i_n}) = F_n(x_1, \ldots, x_n; t_1, \ldots, t_n).$$

An appeal to the Kolmogorov-Daniell extention theorem [cf. Yeh (1973), p. 14] completes the proof of the theorem.

It is possible to describe the probability space $(\Omega, \mathbb{F}, P)$ of the previous theorem more explicitely. We do this now and refer to Yeh (1973), p. 14-17 for details.

Let $S[0, \infty)$ be the collection of all functions $x(.)$ on $[0, \infty)$ to S. Let $D = \{t_1, \ldots, t_n\}$, $0 \leq t_1 < \ldots < t_n$, be a finite subset of $[0, \infty)$. Define

$$R_D(x(.)) = (x(t_1), \ldots, x(t_n))$$

which is a function on $S[0, \infty)$ to S^n. The inverse image

$$R_D^{-1}(B) = \{x(.) \mid x(.) \in S[0, \infty), \ (x(t_1), \ldots, x(t_n)) \in B\}$$

of any subset $B \in S^{(n)}$, is called a Borel cylinder with index D and base B. Let $\mathbf{T}_D = \{R_D^{-1}(B), B \in S^{(n)}\}$ be the collection of all such Borel cylinders with the same index set D. Since $\mathbf{T}_D$ is the inverse image of the power set of S^n, it is a σ-field of subsets of $S[0, \infty)$. The union $\mathbf{T} = \cup \mathbf{T}_D$ of $\mathbf{T}_D$ over all finite subsets D of $[0, \infty)$ is a field and let $\sigma\{\mathbf{T}\}$ denote the minimal σ-field generated by it.

Define Ψ_D on $\mathbf{T}_D$ by

$$\Psi_D(B) = \sum_{(i_1,\ldots,i_n)\varepsilon B} \Psi_n(i_1,\ldots,i_n;\ t_1,\ldots,t_n),\ B \ \varepsilon\ \mathbf{T}_D\ ,$$

where Ψ_n is defined by (14). It can be easily seen to be a probability measure on $\mathbf{T}_D$. One can also verify that Ψ_D defines a probability measure on the field $\mathbf{T}$. Let Ψ denote its unique Carathéodory extention [cf. Loève (1968) p. 87-90] to $\sigma\{\mathbf{T}\}$. Define $X(t, .)$ on $S[0, \infty)$ by

$$X(t, x(.)) = x(t),\quad x(.)\ \varepsilon\ S[0, \infty),\quad t\ \varepsilon\ \mathbb{R}^+ .$$

It is now possible to demonstrate that $(S[0, \infty), \sigma\{\mathbf{T}\}, \Psi)$ is the required probability space and that

$$\{X(t, x(.)),\ t\ \varepsilon\ \mathbb{R}^+),\quad x(.)\ \varepsilon\ S[0, \infty)\}$$

is the sought-after discrete Markov process.

CHAPTER II

THE TRANSITION PROBABILITY FUNCTION

1. INTRODUCTION

We have seen in section 1.6 that the initial distribution and the transition probability function (t.p.f.) of a discrete Markov process determine its finite dimensional distributions. The initial distribution is determined by either a set $\{a_i, i \in S\}$ of arbitrary, non-negative numbers such that

$$\sum_{i \in S} a_i = 1 ,$$

or by the invariant distribution [cf. section 3.6] of the process and which in turn is determined by the t.p.f. It is, therefore, important that we are able to specify the t.p.f. of a finite Markov process before we undertake any further study of its properties.

If we are interested in formulating a finite Markov process as a stochastic model for a real life phenomenon, and if we are required to specify the t.p.f. apriori, we would have no reasonably simple way of accomplishing this task. Moreover, such an apriori specification of the t.p.f. is not in accordance with a basic principle of model-building that the number and nature of assumptions should be few and simple. Fortunately, it is possible to avoid these difficulties by establishing the fact that, under certain conditions, the t.p.f. is determined by the so-called intensity rates.

In section 2, we obtain certain analytic properties of the transition probabilities viewed as functions defined on $\mathbb{R}^+$ and as a consequence establish (i) the existence of the intensity rates and (ii) two differential equations satisfied by the t.p.f. The converse problem of obtaining the t.p.f. as a solution of the differential equations in terms of specified intensity rates is solved in section 3. These results are illustrated in section 4 with the help of the birth-death processes which constitute an important sub-class of finite Markov processes. The last section 5 deals with the properties of the sample functions of a finite Markov process. This enables us to describe a method of simulating a finite Markov process.

2. ANALYTIC PROPERTIES

Let $\{X(t),\ t \in \mathbb{R}^+\}$ be a finite Markov process with state - space S and transition probabilities $p_{jk}(t),\ j,\ k \in S,\ t \in \mathbb{R}^+$ such that

$$p_{jk}(t) \geq 0 \ , \quad \sum_{k \in S} p_{jk}(t) = 1 \ , \qquad 1(i)$$

$$p_{jk}(t+u) = \sum_{r \in S} p_{jr}(t)\, p_{rk}(u)$$

$$= \sum_{r \in S} p_{jr}(u)\, p_{rk}(t) \ ; \qquad 1(ii)$$

and

$$\lim_{t \downarrow 0} p_{jk}(t) = \delta_{jk} = \begin{cases} 1 \ , & \text{if } j = k \ , \\ 0 \ , & \text{otherwise} \ . \end{cases} \qquad 1(iii)$$

We have already encountered 1(i) and 1(ii) in section 1.6, 1(i) being the natural condition on the transition probabilities and 1(ii) being the Chapman-Kolmogorov equations. The only additional restriction is 1(iii) which is equivalent to asserting that $p_{jk}(.)$ is continuous from the right in view of the convention introduced in section 1.3 immediately before Example 1.3.1.

Definition 1 : The transition probabilities $p_{jk}(.)$, $j, k \varepsilon S$ are said to be regular or standard if they satisfy 1(i), 1(ii) and 1(iii).

Hereinafter we shall deal with regular transition probabilities only.

Lemma 1 : The function $p_{jk}(.)$ is uniformly continuous on $[0, \infty)$.

Proof : Let $t \varepsilon IR^+$ and $h > 0$. The Chapman-Kolmogorov equations 1(ii) imply that

$$p_{jk}(t+h) - p_{jk}(t) = -\{1 - p_{jj}(h)\} p_{jk}(t) + \sum_{r \neq j} p_{jr}(h)\, p_{rk}(t)$$

from which it readily follows that

$$|p_{jk}(t+h) - p_{jk}(t)| \leq 1 - p_{jj}(h) \ .$$

The assertion of the lemma is an immediate consequence of 1(iii).

Lemma 2 : The transition probability $p_{jk}(t)$, $j, k \varepsilon S$, is either positive for all $t > 0$ or is zero for all $t > 0$.

Proof : Let $j = k$. Then $p_{jj}(0) = 1$ and 1(iii) implies that for every $t > 0$, $0 < \varepsilon < 1$, there exists an $n_o \varepsilon Z^+$, such that for all $n \geq n_o$, $p_{jj}(t/n) > \varepsilon$. A repeated use of (1(ii)) implies that

$$p_{jj}(t) \geq \{p_{jj}(t/n)\}^n > \varepsilon^n > 0$$

for all $t \geq 0$.

Let $j \neq k$ and suppose $p_{jk}(t_o) > 0$. Using (1(ii)) for $t > t_o$, we find

$$p_{jk}(t) \geq p_{jk}(t_o)\, p_{kk}(t - t_o) > 0.$$

This proves that $p_{jk}(t) > 0$ for $t > t_o$. Suppose, $t \leq t_o$. Then,

$$p_{jk}(t) \geq p_{jk}(u)\, p_{kk}(t-u) \quad \text{for all } u,\ 0 < u \leq t.$$

Therefore, it is enough to demonstrate the existence of an $u \in (0,t)$ such that $p_{jk}(u) > 0$. Let $h = t_o/m$ where m is a positive integer exceeding the finite number M of states in S. A repeated use of 1(ii) yields

$$p_{jk}(t_o) = p_{jk}(mh) = \Sigma p_{jr_1}(h)\, p_{r_1r_2}(h) \ldots p_{r_{m-1}k}(h) \quad ,$$

where the sum extends over all $r_1, \ldots, r_{m-1} \in S$. Since $p_{jk}(t_o) > 0$, there must exist $\nu \leq (M-2)$ distinct states $r_1, \ldots, r_\nu$ such that

$$p_{jr_1}(h)\, p_{r_1r_2}(h) \ \ldots \ p_{r_\nu k}(h) > 0.$$

Now choose $m > \max\{(\nu+1)t_o/t,\ M\}$ and $u = (\nu+1)t_o/m$ so that $u \in (0, t)$ and

$$p_{jk}(u) \geq p_{jr_1}(h)\, p_{r_1 r_2}(h) \ldots p_{r_\nu k}(h) > 0 \; .$$

The proof is complete.

The next natural question is to ask whether the continuous function $p_{jk}(\cdot)$ on $[0, \infty)$ is differentiable. We need the following definition to answer this question.

Definition 2 : The sequence $\{X(nh),\ n \in Z^+\}$, $h > 0$, extracted out of a continuous parameter stochastic process $\{X(t),\ t \in \mathbb{R}^+\}$ is called its discrete skeleton to scale h.

Lemma 3 : The skeleton $\{X(nh),\ n \in Z^+\}$ of a discrete Markov process $\{X(t),\ t \in \mathbb{R}^+\}$ with state-space S is a Markov chain on S. Its n-step transition probabilities are $p_{jk}(nh)$, $j, k \in S$, $n \in Z^+$.

Theorem 1 : For all $j, k \in S$, $j = k$,

$$q_{jk} = \lim_{t \downarrow 0} p_{jk}(t)/t$$

exists and is finite.

Proof : Let $h > 0$ and let $\{X(nh),\ n \in Z^+\}$ be the skeleton to scale h of the discrete Markov process $\{X(t),\ t \in \mathbb{R}^+\}$. Define the taboo probabilities [cf. Chung (1967), p. 45, 46] for n=1,2,...

$${}_k p_{jj}^{(n)}(h) = \Pr[X(nh) = j,\ X(rh) \neq k,\ r = 1, \ldots, (n-1) \,|\, X(0) = j\,],$$

$$f_{jk}^{(n)}(h) = \Pr[X(nh) = k,\ X(rh) \neq k,\ r = 1, \ldots, (n-1) \,|\, X(0) = j]$$

for the skeleton chain $\{X(nh),\ n \in Z^+\}$.

Since $p_{jj}(t)$ and $p_{kk}(t)$ both tend to one and $p_{jk}(t) \to 0$ as $t \downarrow 0$, given ε, $0 < \varepsilon < 1/3$, there exists a $t(\varepsilon)$ such that

$$1 - p_{jj}(t) < \varepsilon, \quad 1 - p_{kk}(t) < \varepsilon, \quad p_{jk}(t) < \varepsilon$$

for all $t \in (0, t(\varepsilon))$. Choose n and h such that $nh < t(\varepsilon)$ and let $u \in [nh, t(\varepsilon)]$. A simple probability argument yields

$$\varepsilon > p_{jk}(u) \geq \sum_{r=1}^{n} f_{jk}^{(r)}(h)\, p_{kk}(r-rh)$$

$$\geq (1-\varepsilon) \sum_{r=1}^{n} f_{jk}^{(r)}(h),$$

so that

$$\sum_{r=1}^{n} f_{jk}^{(r)}(h) \leq \varepsilon/(1-\varepsilon). \tag{2}$$

Partition the event $[X(rh) = j]$ into r disjoint events representing a visit to j at epoch rh with a previous first visit to state k at epoch ℓh, $\ell = 1, \ldots, (r-1)$ and a visit to j at epoch rh without a previous visit to k at any of the epoch ℓh, $\ell = 1, \ldots, (r-1)$. This partitioning yields

$$p_{jj}(rh) = {}_{k}p_{jj}^{(r)}(h) + \sum_{\ell=1}^{(r-1)} f_{jk}^{(\ell)}(h)\, p_{kj}\{(r-\ell)h\}. \tag{3}$$

It follows from (2) and (3) that, for $r \leq n$,

$${}_{k}p_{jj}^{(r)}(h) \geq p_{jj}(rh) - \sum_{\ell=1}^{(r-1)} f_{jk}^{(\ell)}(h)$$

$$\geq (1-\varepsilon) - \varepsilon/(1-\varepsilon) \geq (1-3\varepsilon)/(1-\varepsilon). \tag{4}$$

One more easy probability argument and use of (4) yield

$$p_{jk}(u) \geq \sum_{r=0}^{n-1} {}_k p_{jj}^{(r)}(h)\, p_{jk}(h)\, p_{kk}\{u - (n-r-1)h\}$$

$$> \quad n(1 - 3\varepsilon)\, p_{jk}(h) \;. \tag{5}$$

Now choose n and h such that $nh \leq u \leq (n+1)h$, so that $n \geq (u-h)/h$. Using (5) one has

$$p_{jk}(u)/(u-h) \;\geq\; (1 - 3\varepsilon)\, p_{jk}(h)/h \;. \tag{6}$$

It follows that for all positive $\varepsilon < 1/3$ and $u < t(\varepsilon)$,

$$\limsup_{h \downarrow 0}\; p_{jk}(h)/h \leq p_{jk}(u)/\{u(1-3\varepsilon)\} \;<\; \infty \;. \tag{7}$$

But (6) implies that

$$\limsup_{h \downarrow 0}\; p_{jk}(h)/h \;\leq (1 - 3\varepsilon)\; \liminf_{h \downarrow 0}\; p_{jk}(h)/h$$

which together with (7) yields the result of the theorem by virtue of the arbitrary nature of ε in [0, 1/3].

<u>Corollary</u> : For every state j of a finite Markov process, $[1 - p_{jj}(t)]/t$ converges, as $t \downarrow 0$, to a finite limit $q_j = \sum_{k \neq j} q_{jk}$.

<u>Remark 1</u> : The result of this corollary is not necessarily true for a discrete Markov process with a countable infinity of states. All that one can then assert is that q_j exists but it may be $+\infty$.

2. The above theorem and its corollary establish that the

transition probabilities $p_{jk}(t)$ of a finite Markov process have a right derivative at $t = 0$. One can in fact show that $p_{jk}(.)$ is continuously differentiable at every $t \in [0, \infty)$, [cf. Chung (1967), p. 135]. We shall obtain the derivative $dp_{jk}(t)/dt$ in the next theorem.

3. The stationarity of the transition probabilities enables us to write

$$p_{jk}(h) = p_{jk}(t, t+h) = q_{jk}h + o(h), \quad j \neq k$$

$$p_{jj}(h) = p_{jj}(t, t+h) = 1 - q_j h + o(h) ,$$

for all $t \in \mathbb{R}^+$. Hence the probability of a transition from j to k, $j \neq k$, during $(t, t+h)$, given that $X(t) = j$, and that a change of state has occured, is

$$p_{jk}(h)/\{1 - p_{jj}(h)\} = \{q_{jk}h + o(h)\}/\{q_j h + o(h)\}$$

$$\rightarrow \quad q_{jk}/q_j$$

as $h \downarrow 0$, provided $q_j > 0$. One may thus interpret q_{jk}/q_j, $q_j > 0$, as the conditional probability of a change of state from j to k, given that a transition has occured. It is for this reason that q_{jk}, $j \neq k$, are called the <u>intensity rates</u> of the process.

4. Define

$$r_{jk} = \begin{cases} (1 - \delta_{jk})\, q_{jk}/q_j , & q_j > 0 , \\ \delta_{jk} , & q_j = 0,\ j,\ k \in S, \end{cases}$$

and observe that the matrix $R = ((r_{jk}))$ is a stochastic matrix. It is known as the jump matrix associated with the discrete Markov process $\{X(t),\ t \in \mathbb{R}^+\}$ whose intensity rates are q_{jk}, $j, k \in S$.

Theorem 2 : The transition probabilities $p_{jk}(t)$, $j, k \in S$ of a finite Markov process satisfy the following differential equations :

$$dp_{jk}(t)/dt = \sum_{r \in S} p_{jr}(t)\, q_{rk} \tag{8}$$

$$= \sum_{r \in S} q_{jr}\, p_{rk}(t)\ , \tag{9}$$

where $q_{jj} = -q_j$, $j \in S$, and $t \in (0, \infty)$.

Proof : Let $t > 0$, $h > 0$ and use the Chapman-Kolmogorov equations 1(ii) to obtain

$$\{p_{jk}(t+h) - p_{jk}(t)\}/h$$

$$= -p_{jk}(t)\ \{1 - p_{kk}(h)\}/h + \sum_{r \neq k} p_{jr}(t)\ p_{rk}(h)/h\ . \tag{10}$$

The theorem 1, its corollary and finiteness of S imply that the right side of (10) converges as $h \downarrow 0$, to

$$-q_k p_{jk}(t) + \sum_{r \neq k} p_{jr}(t)\, q_{rk} = \sum_{r \in S} p_{jr}(t)\, q_{rk}$$

with the convention that $q_{kk} = -q_k$, $k \in S$. Hence the left side of (10) must also converge, as $h \downarrow 0$. In fact, it converges to the right derivative of $p_{jk}(t)$ at t. However, since $p_{jk}(.)$ is known to be differentiable on $(0, \infty)$, we have established (8).

The equation (9) follows by an argument similar to the above on using the representation

$$p_{jk}(t+h) = \sum_{r \varepsilon S} p_{jr}(h)\, p_{rk}(t)$$

obtainable from 1(ii). The theorem is established.

Remark 5 : The equations (8) and (9) are respectively known as the forward and the backward equations.

6. Let $S = \{1, 2, \ldots, M\}$ and let Q be the M-square matrix with element q_{jk} in the (j, k) position, j, k =1,..., M . The matrix Q has non-negative entries in off-diagonal ($j \neq k$) positions and non-positive entries along the diagonal such that for every $j = 1, \ldots, M$, $\sum_{k=1}^{M} q_{jk} = 0$. Such a matrix is known as an intensity matrix. If we denote by $dP(t)/dt$ the matrix whose (j,k) entry is $dp_{jk}(t)/dt$, then the forward and backward equations (8) and (9) become, in matrix notation ,

$$dP(t)/dt = P(t)\,Q \tag{11}$$

$$= QP(t)\ . \tag{12}$$

3. SOLUTION OF THE FORWARD AND BACKWARD EQUATIONS

In almost all the problems of Applied Probability, a stochastic model based on a discrete Markov process is specified in terms of the intensity rates q_{jk} $j, k \varepsilon S$. [cf. example 1.3.3] . Thus, in order to determine the transition probabilities one seeks a common unique solution $P(t)$, $t \varepsilon \mathbb{R}^+$ of the forward and backward equations (2.11) and (2.12) respectively, satisfying the initial condition

$$P(0) = I_M \tag{1}$$

and the Chapman-Kolmogorov equation

$$P(u+t) = P(u)\,P(t) = P(t)\,P(u)\ . \tag{2}$$

One, of course, also requires that for each $t > 0$, $P(t)$ be a stochastic matrix. We need the following definition and lemmas to solve the problem posed above.

Definition 1 : Let $A = ((\,a_{jk}))$ be an M-square matrix with real or complex elements. The norm $||A||$ and the exponential matrix $\exp(A)$ of A are defined respectively by

$$||A|| = \sum_{j=1}^{M} \sum_{k=1}^{M} |a_{jk}|\ ,$$

$$\exp(A) = \sum_{n=0}^{\infty} A^n/n!\ ,\quad A^0 = I_M$$

provided the matrix series on the right converges element-wise.

Lemma 1 : If A and B are two M-square matrices then

$$||AB|| \le ||A||\ \ ||B||$$

Proof : This is an immediate consequence of the following string of elementary inequalities :

$$||AB|| = \sum_{j=1}^{M} \sum_{k=1}^{M} \Big| \sum_{r=1}^{M} a_{jr}\, b_{rk} \Big|$$

$$\leq \sum_j \sum_k \sum_r |a_{jr} b_{rk}|$$

$$\leq \sum_j \sum_k |a_{jk}| \sum_\ell \sum_m |b_{\ell m}|$$

$$= \|A\| \ \|B\| .$$

Lemma 2 : If $\{A_n, n \in Z^+\}$ is a sequence of M-square matrices such that $\sum_{n=0}^{\infty} \|A_n\| < \infty$, then $\sum_{n=0}^{\infty} A_n$ converges in the sense that all the M^2 series $\sum_{n=0}^{\infty} a_{jk}^{(n)}$ converge, where $a_{nk}^{(n)}$ is the (j, k) element of A_n, j, k = 1, ..., M .

Lemma 3 : The exponential matrix exp(A) of any M-square matrix is well defined.

Proof : This is an immediate consequence of lemma 2 and the fact that by lemma 1

$$\sum_{n=0}^{\infty} \|A^n\|/n! \leq \sum_{n=0}^{\infty} \|A\|^n/n! < \infty .$$

Lemma 4 : If A and B are M-square matrices which commute, i.e. if AB = BA, then

$$\exp(A+B) = \exp(A)\exp(B) = \exp(B)\exp(A) .$$

Lemma 5 : If A is an M-square matrix, then for all $t \in [0, \infty)$,

$$de^{tA}/dt = A \exp(tA) = \exp(tA) A,$$

where the differentiation is elementwise.

Proof : The (j, k) element of exp (tA) is

$$\sum_{n=0}^{\infty} t^n a_{jk}^{(n)}/n !$$

where $a_{jk}^{(n)}$ is the (j, k) element in A^n. This series is uniformly convergent for $t \in [0, \infty)$ as it is dominated in absolute value by the uniformly convergent series $\sum_{n=0}^{\infty} t^n \|A\|^n/n!$. It follows that

$$d\{\sum_{n=0}^{\infty} t^n a_{jk}^{(n)}/n!\}/dt = \sum_{n=0}^{\infty} t^n a_{jk}^{(n+1)}/n! .$$

The result of the lemma becomes obvious if we note that for all j, k = 1, 2, ..., M

$$a_{jk}^{(n+1)} = \sum_{r=1}^{M} a_{jr} a_{rk}^{(n)} = \sum_{r=1}^{M} a_{jr}^{(n)} a_{rk} .$$

Lemma 6 : Let $A(t)$, $t \in \mathbb{R}^+$, be M-square matrices with elements $a_{jk}(t)$ which are continuous integrable functions. Then for every $t > 0$,

$$\left\| \int_0^t A(u)\, du \right\| \leq \int_0^t \|A(u)\| du ,$$

where the first integration is elementwise.

Proof : This is an immediate consequence of the definition of $\|A(t)\|$ and elementary properties of Riemann integrals.

Theorem 1 : The forward and backward equations (2.11)and (2.12) have the common unique solution

$$P(t) = \exp(tQ), \quad t \in \mathbb{R}^+, \tag{3}$$

which satisfies the initial condition (1) and the Chapman-Kolmogorov equation (2). Moreover, the matrix $P(t)$ defined by (3) is a stochastic matrix for each $t \in \mathbb{R}^+$.

Proof : The lemma 5 implies that (3) is a solution of both (2.11) and (2.12) satisfying the initial condition (1). We claim that it is the only solution of (2.11) and (2.12).

Let, if possible, $P_1(t)$ and $P_2(t)$ be two common solutions of (2.11) and (2.12) satisfying (1). We can rewrite equation (2.12) as

$$P(t) = I_M + Q \int_0^t P(u)\, du$$

from which it follows that

$$P_1(t) - P_2(t) = Q \int_0^t \{P_1(u) - P_2(u)\}\, du. \tag{4}$$

Hence by lemmas 1 and 6,

$$\|P_1(t) - P_2(t)\| \leq \|Q\| \int_0^t \|P_1(u) - P_2(u)\|\, du. \tag{5}$$

Since the elements of $P_1(t)$ and $P_2(t)$ are differentiable and therefore continuous functions of t, we must have

$$0 \leq \alpha = \sup \{\|P_1(u) - P_2(u)\| \mid 0 \leq u \leq t < \infty\} < \infty.$$

It follows from (5) that

$$\|P_1(t) - P_2(t)\| \leq \alpha \|Q\| t$$

using which on the right of (5) one has

$$\|P_1(t) - P_2(t)\| \leq \alpha^2 t^2 \|Q\|/2\,!$$

A simple induction argument yields

$$\|P_1(t) - P_2(t)\| \leq \alpha^n t^n \|Q\|^n/n! \to 0$$

as $n \to \infty$. It follows that $P_1(t) \equiv P_2(t)$, $t \in [0, \infty)$.

The fact that the common unique solution (3) satisfies the Chapman-Kolmogorov equation is an easy consequence of lemma 4.

It now remains to demonstrate that $\exp(tQ)$ is a stochastic matrix. Observe that for any real number c, the matrix $\exp(-ctI_M)$ has non-negative elements. Choose $c > \max(q_1, \ldots, q_M)$ so that $Q + cI_M$ and therefore $\exp(tQ + ctI_M)$ have non-negative elements for every $t \in \mathbb{R}^+$. The non-negative nature of elements of $\exp(tQ)$ follows from the obvious identity

$$\exp(tQ) = \exp(tQ + ct\, I_M)\exp(-ct\, I_M).$$

If E_{M1} denotes the Mx1 column vector with all elements equal to 1, by definition, it follows that $Q\,E_{M1}$, is the Mx1 null vector. Hence so is $Q^n E_{M1}$ for all $n \geq 1$. Thus

$$\exp(tQ)\,E_{M1} = I_M\,E_{M1} + \sum_{n=1}^{\infty} t^n Q^n\,E_{M1}/n! = E_{M1},$$

establishing that $\exp(tQ)$ is a stochastic matrix for all $t \in \mathbb{R}^+$. The proof is complete.

An alternative form of the solution of the forward and backward equations is provided by the following

Theorem 2 : Let Q be an M-square intensity matrix, $\beta > \max(q_1, \ldots, q_M)$ and $P = I + \beta^{-1} Q$. Then

$$P(t) = \exp(tQ) = \exp(-\beta t) \sum_{n=0}^{\infty} (\beta t)^n P^n/n! \ , \ t \in \mathbb{R}^+ . \quad (6)$$

Proof : Observe that, by construction, P is a stochastic matrix. Moreover,

$$\exp[-\beta t] \sum_{n=0}^{\infty} (\beta t)^n P^n/n!$$

$$= \exp(\beta t) \sum_{n=0}^{\infty} [\{(\beta t)^n/n!\} \sum_{k=0}^{n} \binom{n}{k} Q^k/\beta^k]$$

$$= \exp(-\beta t) \sum_{k=0}^{\infty} \{t^k Q^k/k!\} \sum_{n=k}^{\infty} (\beta t)^{n-k}/(n-k)!$$

$$= \exp(tQ) ,$$

which concludes the proof by virtue of (3) and the uniqueness of the solution established in theorem 1.

Remark : We have already seen in lemma 1.3.2 that the process derived from a Markov chain, using an independent Poisson process as the deriving process, is a discrete Markov process. The above theorem asserts that every finite Markov process is equivalent to a process derived from a Markov chain by an independent Poisson process, in the sense that they have the same family of finite dimensional distributions.

We now describe an iterative form of the solution of the forward and the backward equations. The equations (2.8) and (2.9) can be rewritten in the following forms :

$$P_{jk}(t) = \exp(-q_k t)\,\{\delta_{jk} + \int_0^t \sum_{r \neq k} (p_{jr}(u)\, q_{rk}) \exp(q_k u) du\} \tag{7}$$

$$= \exp(-q_j t)\,\{\delta_{jk} + \int_0^t \sum_{r \neq j} (q_{jr} p_{rk}(u)) \exp(q_j u)\, du\}\,, \tag{8}$$

where δ_{jk} is the Kronecker delta.

Let $Q_{dg} = \text{diag}\,(-q_1, \ldots, -q_M)$ be the M-square diagonal matrix with elements $-q_1, \ldots, -q_M$ along the main diagonal and define

$$H(t) = \exp\,(t\, Q_{dg}) \quad .$$

The M^2 equations (7) and (8), when represented in matrix notation, become respectively

$$P(t) = [I_M + \int_0^t P(u)\,(Q - Q_{dg})\, H(u)\, du]\, H(-t) \tag{9}$$

$$= H(-t)\,[I_M + \int_0^t H(u)\,(Q - Q_{dg})\, P(u)\, du]\,. \tag{10}$$

Use the expression provided by (10) for $P(u)$ in (10) itself to obtain

$$P(t) = H(-t) + H(-t) \int_0^t H(u)\,(Q - Q_{dg})$$

$$\{H(-u) + H(-u) \int_0^u H(v)\,(Q - Q_{dg})\, P(v)\, dv\}\, du$$

$$= H(-t) + H(-t) \int_0^t H(u)\,(Q - Q_{dg})\, P(u)\, du$$

$$+ H(-t) \int_0^t H(u)(Q-Q_{dg})\, H(-u)\{\int_0^u H(v)(Q-Q_{dg})P(v)dv\}du\,.$$

Repeating the argument one can finally obtain

$$P(t) = \sum_{n=0}^{\infty} P_n(t) , \qquad (11)$$

where $P_o(t) = H(-t)$ and

$$P_{n+1}(t) = H(-t) \int_0^t H(u)\,(Q-Q_{dg})\,P_n(u)\,du, \; n \in Z^+, \; t \in \mathbb{R}^+ .$$

In an analogous manner equation (9) can be used to obtain

$$P(t) = \sum_{n=0}^{\infty} P_n^*(t) \qquad (12)$$

where $P_o^*(t) = H(-t)$ and

$$P_{n+1}^*(t) = \{ \int_0^t P^*(u)\,(Q-Q_{dg})\,H(u)\,du\}\,H(-t), \; n \in Z^+, \; t \in \mathbb{R}^+ .$$

The equations (11) and (12) have been obtained in a formal manner but one can use the theory of integral equations to demonstrate that $\Sigma P_n(t)$, $\Sigma P_n^*(t)$, converge for all $t>0$, and are common solutions of the forward and backward equations with all the necessary properties.

4. EVALUATION OF THE TRANSITION PROBABILITIES.

This section is devoted to a description of the methods of evaluating the exponential matrix $\exp(tQ)$ where Q is the intensity matrix of a finite Markov process with state-space $S = \{1, 2, \ldots, M\}$. By definition

$$\exp(tQ) = \sum_{n=0}^{\infty} t^n Q^n/n! \qquad (1)$$

and therefore the main problem is that of obtaining the matrix Q^n and then of summing the infinite series (1). The following lemma shows that this is a simple matter if all the characteristic roots of Q are distinct.

Lemma 1 : Let $\alpha_1, \ldots, \alpha_M$ denote the M distinct characteristic roots of Q and let ξ_r and η_r denote the $M\times1$ right and left characteristic vectors of Q corresponding to the root α_r, $r = 1, \ldots, M$, so normalized that

$$\xi_r^T \eta_s = \delta_{rs} \text{ , the Kronecker delta.}$$

Then

$$P(t) = \exp(tQ) = \sum_{r=1}^{M} \exp(\alpha_r t)\, \xi_r\, \eta_r^T . \qquad (2)$$

Proof : By definition

$$Q\xi_r = \alpha_r \xi_r \ , \quad \eta_r^T Q = \alpha_r \eta_r^T \ , \quad r = 1, \ldots, M .$$

The method of spectral resolution described in example 1.3.2 implies that

$$Q^n = \sum_{r=1}^{M} \alpha_r^n \xi_r \eta_r^T$$

of which (2) is an easy consequence.

The use of expression (2) to obtain the transition probabilities requires us to calculate the characteristic roots and characteristic vectors of Q. We, therefore, proceed to discuss the properties of the characteristic roots and vectors of Q which are also of in - dependent interest.

Lemma 2 : Zero is a characteristic root of an intensity matrix Q and the real part of every characteristic root of Q is non - positive.

Proof : The first assertion is an immediate consequence of the fact that if Q is an intensity matrix and E_{M1} is the Mx1 column vector with all elements equal to 1, then $QE_{M1} = 0$.

Let α be a characteristic root of Q and let $\xi = (x_1,\ldots,x_M)^T$ be the corresponding right characteristic vector. Then, by definition,

$$\sum_{r=1}^{M} (q_{jr} - \alpha\delta_{jr})\, x_r = 0 , \qquad j = 1, 2, \ldots, M,$$

so that

$$|q_j + \alpha|\,|x_j| = |\sum_{r\neq j} q_{jr}\, x_r| \leq \sum_{r\neq j} q_{jr}|\, x_r| . \tag{3}$$

Let $\beta = \max(|x_1|,\ldots, |x_M|) = |x_\nu|$ say. It follows from (3) with $j = \nu$, that

$$|q_\nu + \alpha| \leq q_\nu\, \beta \cdot$$

Thus α, which is possibly a complex number, lies within or on the circle with centre at $-q_\nu$ and radius q_ν . It follows that the real part of α can not be positive. The proof is complete.

Lemma 3 : If α is a characteristic root of Q and ξ and η are its corresponding right and left characteristic vectors, then exp (αt) is a characteristic root of P(t) and ξ and η are its corresponding right and left characteristic vectors respectively.

Proof : Since

$$Q\xi = \alpha\xi , \quad \eta^T Q = \eta^T \alpha ,$$

we have

$$Q^n \xi = \alpha^n \xi, \quad \eta^T Q^n = \alpha^n \eta^T$$

for all $n \geq 1$. The lemma follows immediately.

We now describe a method of obtaining the transition probabilities corresponding to an intensity matrix Q which has only real characteristic roots, distinct or otherwise, without being required to obtain the right and left characteristic vectors explicitely. These results are due to Chiang and Raman (1971).

In what follows we shall denote by $\Delta(\alpha)$ the determinant $|\alpha I_M - Q|$ and by $\varphi(\alpha)$ the adjoint matrix $((A_{jk}(\alpha)))$ of $\alpha I_M - Q$, where the (j.k)-element $A_{jk}(\alpha)$ of $\varphi(\alpha)$ is the co-factor of the (k.j)-element $\alpha\delta_{kj} - q_{kj}$ of $\alpha I_M - Q$. The following simple lemmas are needed in later development.

Lemma 4 : If $x_1, \ldots, x_M$ are any M distinct real numbers, then

$$\sum_{j=1}^{M} x_j^r / \{ \prod_{k \neq j} (x_j - x_k) \} = \begin{cases} 0 , & \text{if } 0 \leq r < M-1, \\ 1 , & \text{if } r = M-1 , \end{cases}$$

where (and in what follows) $\prod_{k \neq j} (x_j - x_k)$ stands for

$$\prod_{\substack{k=1 \\ k \neq j}}^{M} (x_j - x_k) .$$

Proof : [Cf. Chiang (1968), p. 126] .

Lemma 5 : Let $\varphi^{(r)}(\alpha) = d^r\varphi(\alpha)/d\alpha^r$, $\varphi^{(o)}(\alpha) = \varphi(\alpha)$, and $\Delta^{(r)}(\alpha) = d^r\ \Delta(\alpha)/d\alpha^r$, $r = 1, 2, \ldots$. Then

$$\alpha\varphi^{(r)}(\alpha) = Q\varphi^{(r)}(\alpha) - r\varphi^{(r-1)}(\alpha) + \Delta^{(r)}(\alpha)\ I_M , \tag{4}$$

and if α^* is a root of multiplicity $m \geq 1$ of $\Delta(\alpha) = 0$ then

$$\Delta^r(\alpha^*) = [d^r\varphi(\alpha)/d\alpha^r]_{\alpha=\alpha^*} = 0, \quad r \leq m-1 .$$

Proof : The relation (4) is an easy consequence of the well-known relation

$$(\alpha I_M - Q)\ \varphi(\alpha) = \Delta(\alpha)\ I_M \tag{5}$$

and can be established by induction. The second assertion is an immediate consequence of the multiplicity of α^* .

Theorem 1 : Suppose that the intensity matrix has distinct and real characteristic roots $\alpha_1, \ldots, \alpha_M$. Then

$$p_{jk}(t) = \sum_{r=1}^{M} \exp(\alpha_r t)\ A_{jk}(\alpha_r) / \prod_{\ell \neq r} (\alpha_r - \alpha_\ell),\ j, k = 1,\ldots,M . \tag{6}$$

Proof : Observe that for each α_r, every column of $\varphi(\alpha_r)$ as well as of the matrix $\xi_r\eta_r^T$, where ξ_r and η_r are respectively the right and left characteristic vectors of Q corresponding to the root α_r, is a solution of the system

$$(\alpha_r\ I_M - Q)\xi = 0 \tag{7}$$

of homogeneous linear equations. Since α_r is a simple root, the system (7) has only one linearly independent solution. Hence a column of the matrix $\xi_r \eta_r^T$ is proportional to any column of $\varphi(\alpha_r)$, i.e., if $\gamma_{jk}(r)$ is the (j, k) element of $\xi_r \eta_r^T$, we can write

$$\gamma_{jk}(r) = C_k(r) A_{jk}(r) , \quad j = 1, \ldots, M ,$$

where $C_k(r)$ is a constant of proportionality. The representation

$$P(t) = \sum_{r=1}^{M} \exp(\alpha_r t) \xi_r \eta_r^T , \quad P(0) = I_M$$

implies that

$$\sum_{r=1}^{M} \gamma_{jk}(r) = \sum_{r=1}^{M} C_k(r) A_{jk}(r) = \delta_{jk}, \quad j, k = 1,\ldots, M,$$

which can be regarded as a set of M^2 linear equations in the M^2 unknowns $C_k(r)$, r, k = 1, ..., M. Recalling that $A_{jk}(r)$ is a polynomial of degree (M-1) in α_r and using lemma 4, one can prove that

$$C_k(r) = \{ \prod_{\ell \neq r} (\alpha_r - \alpha_\ell)\}^{-1} , \quad r, k = 1,\ldots,M.$$

We thus have (6) and the theorem is proved.

Theorem 2 : Suppose the intensity matrix Q has one real characteristic root α_1 of multiplicity m, $1 < m < M$ and that its remaining roots $\alpha_{m+1}, \ldots, \alpha_M$ are real and of multiplicity one each. Then

$$p_{jk}(t) = \frac{\exp(\alpha_1 t)}{\prod\limits_{\ell=m+1}^{M} (\alpha_1 - \alpha_\ell)} \sum_{u=0}^{m-1} \frac{t^{m-1-u}}{(m-1-u)!\, u!} \frac{d^u}{d\alpha_1{}^u} A_{jk}(\alpha_1)$$

$$+ \sum_{r=m+1}^{M} \exp(\alpha_r t)\, A_{jk}(\alpha_r)/\{(\alpha_r - \alpha_1)^m \prod_{\substack{\ell=m+1 \\ \ell \neq r}}^{M} (\alpha_r - \alpha_\ell)\}. \qquad (8)$$

Proof : Consider the expression on the right of (6) as a function $g(\alpha_1, \ldots, \alpha_M)$ of $\alpha_1, \ldots, \alpha_M$. Replace $\alpha_1, \ldots, \alpha_M$ by $\alpha_1 + \varepsilon_1, \ldots, \alpha_1 + \varepsilon_m$ respectively and leave $\alpha_{m+1}, \ldots, \alpha_m$ unchanged, to obtain

$$g(\alpha_1 + \varepsilon_1, \ldots, \alpha_1 + \varepsilon_m, \alpha_{m+1}, \ldots, \alpha_m)$$

$$= \sum_{r=1}^{m} \exp\,[(\alpha_1 + \varepsilon_r)t]\, A_{jk}(\alpha_1 + \varepsilon_r)/[\{\prod_{\substack{\ell=1 \\ \ell \neq r}}^{m} (\varepsilon_r - \varepsilon_\ell)\}\{\prod_{\ell=m+1}^{M} (\alpha_1 + \varepsilon_r - \alpha_\ell)\}]$$

$$+ \sum_{r=m+1}^{M} \exp(\alpha_r t)\, A_{jk}(\alpha_r)/[\{\prod_{\ell=1}^{m} (\alpha_r - \alpha_1 - \varepsilon_\ell)\}\{\prod_{\ell=m+1}^{M} (\alpha_r - \alpha_\ell)\}]. \qquad (9)$$

Now consider

$$\lim g(\alpha_1 + \varepsilon_1, \ldots, \alpha_m + \varepsilon_m, \alpha_{m+1}, \ldots, \alpha_M)$$

as $\varepsilon_1, \ldots, \varepsilon_m \to 0$. Expand $A_{jk}(\alpha_1 + \varepsilon_r)$ around α_1 by Taylor expansion to obtain

$$\exp[(\alpha_1+\varepsilon_r)t]\, A_{jk}(\alpha_1+\varepsilon_r)$$

$$= \exp(\alpha_1 t) \sum_{n=0}^{\infty} \varepsilon_r^n \{ \sum_{u=0}^{n} \frac{t^{n-u}}{(n-u)!u!} \frac{d^u}{d\alpha_1^u} A_{jk}(\alpha_1)\} .$$

Thus the first term on the right of (9) is

$$\exp(\alpha_1 t) \sum_{n=0}^{\infty} [\sum_{r=1}^{m} \varepsilon_r^n/[\{ \prod_{\substack{\ell=1\\ \ell\neq r}}^{m} (\varepsilon_r-\varepsilon_\ell)\}\{ \prod_{\ell=m+1}^{M} (\alpha_1+\varepsilon_r-\alpha_\ell)\}]]$$

$$\times [\sum_{u=0}^{n} \frac{t^{n-u}}{(n-u)!\, u!} [\frac{d^u}{d\alpha_1^u} A_{jk}(\alpha_1)]] . \qquad (10)$$

It is easy to verify that for $n \geq m$

$$\lim_{\varepsilon_1,\ldots,\varepsilon_m\to 0} \varepsilon_r^n/\{ \prod_{\substack{\ell=1\\ \ell\neq r}}^{m} (\varepsilon_r-\varepsilon_\ell)\} = 0 \qquad (11)$$

and that by lemma 4 ,

$$\sum_{r=1}^{m} \varepsilon_r^n/\{ \prod_{\substack{\ell=1\\ \ell\neq r}}^{m} (\varepsilon_r-\varepsilon_\ell)\} = \begin{cases} 1, & \text{if } n = m-1, \\ 0, & \text{if } 0 \leq n < m-1. \end{cases} \qquad (12)$$

Using (11) and (12) in (10), we find that the limit of the first term on the right in (9), as $\varepsilon_1, \ldots, \varepsilon_m \to 0$, is

$$\frac{\exp(\alpha_1 t)}{\prod_{\ell=m+1}^{M} (\alpha_1-\alpha_\ell)} \sum_{u=0}^{m-1} \frac{t^{m-1-u}}{(m-1-u)!\ u!} \frac{d^u}{d\alpha_1^u} A_{jk}(\alpha_1) . \tag{13}$$

The limit of the second term on the right in (9) is easily seen to be

$$\sum_{r=m+1}^{M} \exp(\alpha_r t) A_{jk}(\alpha_r)/\{(\alpha_r-\alpha_1)^m \prod_{\substack{\ell=m+1\\ \ell\neq r}}^{M} (\alpha_r-\alpha_\ell)\} . \tag{14}$$

Combining (13) and (14), one finds that the function g converges to the expression on the right of (8) as $\varepsilon_1,\ldots,\varepsilon_m \to 0$. It now remains to show that this limit, to be denoted by $g_{jk}(\alpha_1, \ldots, \alpha_1, \alpha_{m+1}, \ldots, \alpha_M)$, is a solution of the forward and backward equations (2.11) and (2.12). Let therefore $G(t)$ denote the matrix $((g_{jk}))$ and observe that

$$G(t) = \lambda \exp(\alpha_1 t) \sum_{u=0}^{m-1} t^{m-1-u} \varphi^{(u)}(\alpha_1)/\{(m-1-u)!\ u!\}$$

$$+ \sum_{r=m+1}^{M} \mu_r \exp(\alpha_r(t), \varphi(\alpha_r)) , \tag{15}$$

where

$$\lambda = \{\prod_{\ell=m+1}^{M} (\alpha_1-\alpha_\ell)\}^{-1} \quad \text{and} \quad \mu_r = \{(\alpha_r-\alpha_1)^m \prod_{\substack{\ell=m+1\\ \ell\neq r}}^{M} (\alpha_r-\alpha_\ell)\}^{-1} .$$

Differentiating both the sides of (15) with respect to t and using lemma 5, it easily follows that $G(t)$ is a solution of

(2.11) and (2.12), so that the proof of the theorem is complete.

Corollary 1 : Suppose the matrix Q has real characteristic roots $\alpha_1, \ldots, \alpha_m$ of multiplicities $\nu_1, \ldots, \nu_m$ such that $\Sigma\nu_j = M$. Then

$$p_{jk}(t) = \sum_{r=1}^{m} \frac{\exp(\alpha_r t)}{\prod_{\substack{\ell=1 \\ \ell\neq r}}^{m} (\alpha_r - \alpha_\ell)} \sum_{u=0}^{\nu_r-1} \frac{t^{\nu_r-1-u}}{(\nu_r-u-1)!u!} \varphi_{jk}^{(u)}(\alpha_r) . \tag{16}$$

Proof : The expression (16) for $p_{jk}(t)$ is obtained by using the procedure adopted in the proof of theorem 2 for $\alpha_2, \ldots, \alpha_m$ successively.

In order to illustrate the techniques developed in this section, we introduce in the next section a special type of finite Markov processes called the finite birth-death processes which are also of independent interest.

5. FINITE BIRTH AND BIRTH-DEATH PROCESSES

This section is devoted to the study of some properties of a sub-class of finite Markov processes known as the finite birth and birth-death process.

Definition 1 : A finite Markov process $\{X(t), t \in \mathbb{R}^+\}$ with state-space $S = \{0, 1, \ldots, M\}$ such that

$$\Pr[X(t+h) = k|X(t) = j] = \begin{cases} \lambda_j h + o(h), & \text{if } k = j+1, 0 \leq j < M, \\ 1 - \lambda_j h + o(h), & \text{if } k = j, 0 \leq j < M, \\ 1, & \text{if } k = j = M, \\ o(h), & \text{otherwise} ; \end{cases} \tag{1}$$

where $\lambda_0, \lambda_1, \ldots, \lambda_{M-1}$ are finite, positive constants, is called a finite birth process.

If we interpret $X(t)$ as the size of a population of individuals at epoch t, then conditional on $X(t) = j$, $0 \leq j < M$, $\lambda_j h+o(h)$ can be looked upon as the probability of a 'birth' during $(t, t+h]$. It is for this reason that $\lambda_0, \lambda_1, \ldots, \lambda_{M-1}$ are referred to as birth rates and the process is referred to as a birth process.

The only positive intensity rates of the birth process are $q_{jj+1} = \lambda_j$ and $q_j = -q_{jj} = \lambda_j$, $0 \leq j < M$. The forward equations (2.8) are :

$$\begin{aligned} dp_{j0}(t)/dt &= -\lambda_0 p_{j0}(t) , \\ dp_{jk}(t)/dt &= -\lambda_k p_{jk}(t) + \lambda_{k-1} p_{jk-1}(t), 1 \leq k \leq M-1, \quad (2) \\ dp_{jM}(t)/dt &= \lambda_{M-1} p_{jM-1}(t) , \end{aligned}$$

which are to be solved subject to the initial conditions $p_{jk}(0) = \delta_{jk}$. The equations (2) are easily seen to have the following recursive form of the solution :

$$p_{j0} = \delta_{j0} \exp(-\lambda_0 t) ,$$

$$p_{jk}(t) = \delta_{jk} \exp(-\lambda_k t) + \exp(-\lambda_k t) \int_0^t \lambda_{k-1} p_{jk-1}(u) \exp(\lambda_k u) du,$$

for $k = 1, \ldots, M-1$, and $p_{jM}(t)$ is determined by the relation

$$p_{jM}(t) = 1 - \sum_{k=0}^{M-1} p_{jk}(t), \quad t \in IR^+ .$$

One can easily verify by induction that when the birth rates $\lambda_0, \ldots, \lambda_{M-1}$ are all distinct,

$$p_{jk}(t) = \begin{cases} 0, & \text{if } k < j, \\ \exp(-\lambda_j t), & k = j, \\ \lambda_j \lambda_{j+1} \cdots \lambda_k \sum_{r=j}^{k} \exp(-\lambda_r t) [\prod_{\ell=j}^{k} (\lambda_\ell - \lambda_r)]^{-1}, & \end{cases} \tag{3}$$

where in the last expression $\ell \neq r$ and $j < k \leq M-1$. The above method can also be used if some or all of the birth rates are equal or alternatively the explicit expressions for $p_{jk}(t)$ can be obtained by a limiting procedure in (3) employing L'Hospital's rule.

Two special cases of interest are :

$$\text{Case (i)} : \lambda_j \equiv \lambda, \quad j = 0, 1, \ldots, M-1$$

$$\text{Case (ii)} : \lambda_j = j\lambda, \quad j = 1, 2, \ldots, M-1, \tag{4}$$

where in case (ii), the state zero is deleted from the state-space for obvious reasons. One can easily verify that in case (i)

$$p_{0k}(t) = (\lambda t)^k \exp(-\lambda t)/k!, \quad k = 0, 1, \ldots, M-1,$$

and

$$p_{0M}(t) = 1-\exp(-\lambda t) \{1+\lambda t + (\lambda t)^2/2!+\ldots+(\lambda t)^{M-1}/(M-1)!\} \tag{5}$$

i.e., the conditional distribution of $X(t)$, given $X(0) = 0$, is the truncated Poisson distribution.

In case (ii), deleting state zero, one has

$$p_{1k}(t) = \exp(-\lambda t)\{1 - \exp(-\lambda t)\}^{k-1}, \quad k = 1,\ldots,M-1,$$

$$p_{1M}(t) = \{1 - \exp(-\lambda t)\}^{M-1} \tag{6}$$

i.e., the conditional distribution of $X(t)$, given $X(0) = 1$, is the truncated geometric distribution.

The case (ii) may be interpreted as follows. Suppose $X(t) = j$ corresponds to existence of j individuals in the population such that each individual, independently of all others, has the same probability $\lambda h + o(h)$ of producing one offspring during $(t, t+h]$, deaths being ruled out as impossible. One, of course, has also to assume that reproduction is impossible after reaching the population size M. Then it is easy to see that $\{X(t), t \in \mathbb{R}^+\}$ is the birth process with $\lambda_j = j\lambda$, $j = 1, \ldots, M-1$.

A simple generalization of the birth process is obtained by allowing transitions out of state j to either of the neighbouring states $j \pm 1$. More specifically, we have

Definition 2 : A Markov process $\{X(t), t \in \mathbb{R}^+\}$ on the state-space $\{0, 1, \ldots, M\}$ is a birth-death process if

$$\Pr[X(t+h) = k|X(t) = j] = \begin{cases} \lambda_j h + o(h), & k = j+1,\ 0 \le j < M, \\ \mu_j h + o(h), & k = j-1,\ 0 < j \le M, \\ 1 - (\lambda_j+\mu_j)h + o(h), & 0 \le j \le M, \\ o(h), & \text{otherwise}; \end{cases} \tag{7}$$

where λ_j and μ_j, $j = 0, 1, \ldots, M$ are all positive constants with the exception of λ_M and μ_o which are zero.

The non-zero intensity rates of a birth-death process are

$$q_{jj+1} = \lambda_j, \quad 0 \leq j \leq M-1, \quad q_{jj-1} = \mu_j \ , \quad 0 < j \leq M$$

and consequently $q_j = -q_{jj} = (\lambda_j + \mu_j), \quad 0 \leq j \leq M.$
The system of forward differential equations is

$$dp_{jk}(t)/dt = -(\lambda_k+\mu_k)p_{jk}(t) + \lambda_{k-1}p_{jk-1}(t) + \mu_{k+1}p_{j(k+1)}(t),$$
$$0 \leq j \leq M \ , \qquad (8)$$

provided we interpret $p_{0(-1)}(t) = P_{M(M+1)}(t) \equiv 0$. It is not possible to provide a simple solution of the type (3) in case of the above system (8) of equations. We, therefore, discuss some specific examples.

<u>Example 1</u> : <u>Random Walk with reflecting barriers</u>.

This is a birth-death process on the state-space $\{0,1,\ldots,M\}$ with $\lambda_j \equiv \lambda$, $0 \leq j < M$ and $\mu_j \equiv \mu$, $0 < j \leq M$. It corresponds to the random motion of a hypothetical particle on the set of integers 0, 1, ..., M, such that if it is at position j at epoch t, then during (t, t+h], it moves to (j+1) with probability $\lambda h + o(h)$, to (j-1) with probability $\mu h + o(h)$ and does not shift its position with probability $1 - (\lambda+\mu)h + o(h)$, $0 < j < M$. The barries at 0 and M are reflecting in the sense that the random walking particle when at zero (at M), either stays their or is reflected back to 1 (to M-1), according to the above mentioned probabilities, with slight but obvious modifications.

The non-zero entries in the (M+1)-square intensity matrix Q are

$$q_{jj+1} = \lambda\ ,\ j = 0,\ 1,\ \ldots,\ M-1,\quad q_{jj-1} = \mu\ ,\ j = 1,\ldots,\ M, \tag{9}$$
$$q_{00} = -\lambda\ ,\ q_{MM} = -\mu,\quad q_{jj} = -(\lambda+\mu),\ 1 \leq j \leq M-1\ .$$

We proceed to obtain the characteristic roots and vectors of the intensity matrix Q.

If α is a characteristic root of Q, then we should be able to find a non-trivial (M+1) x 1 vector $\xi = (x_0, x_1, \ldots, x_M)^T$ such that $Q\xi = \alpha\xi$ or equivalently

$$\mu x_{j-1} - (\lambda+\mu+\alpha)x_j + \lambda x_{j+1} = 0,\quad 0 \leq j \leq M, \tag{10}$$

with the boundary conditions

$$x_o = x_{-1},\quad x_M = x_{M+1}\ . \tag{11}$$

The most general solution of the difference equation (10) is

$$x_j = c_1\theta_1^j + c_2\theta_2^j\ ,\quad j = 0,\ 1,\ \ldots,\ M \tag{12}$$

if the roots θ_1, θ_2 of the auxiliary equation

$$\mu - (\lambda+\mu+\alpha)\theta + \lambda\theta^2 = 0 \tag{13}$$

are distinct and is

$$x_j = (c_1 + jc_2)\ \theta_o^j\ ,\quad j = 0,\ 1,\ \ldots,\ M \tag{14}$$

if $\theta_1 = \theta_2 = \theta_o$. Here c_1 and c_2 are constants to be determined by using the boundary conditions (11).

Observe from (13) that $\theta_1\theta_2 = \mu/\lambda > 0$ so that none of θ_1 and θ_2 can be zero. If either of them is equal to one, the other must be μ/λ and then since $\theta_1 + \theta_2 = (\lambda+\mu+\alpha)/\lambda$, α must be zero, which by lemma 4.2 is a characteristic root of Q. If $\theta_0 \neq 1$, the boundary conditions (11) imply that $c_1 = c_2 = 0$ which leads to the trivial null solution of (10). Thus we are interested only in those values of α which lead to distinct solutions of (13) with none of them equal to one. In this situation the boundary conditions(11) lead to the equations :

$$c_1(\theta_1^{-1} - 1) + c_2(\theta_2^{-1} - 1) = 0 \tag{15}$$

$$c_1\theta_1^{M+1}(\theta_1^{-1} - 1) + c_2\theta_2^{M+1}(\theta_2^{-1} - 1) = 0 ,$$

which imply

$$c_2/c_1 = -(\theta_1^{-1} - 1)/(\theta_2^{-1} - 1) , \tag{16}$$

and

$$\theta_1^{M+1} = \theta_2^{M+1} , \qquad \theta_1 \neq \theta_2 . \tag{17}$$

Since $\theta_1\theta_2 = \mu/\lambda = \rho^2$ say, we must have

$$(\rho^{-1}\theta_1)^{2M+2} = 1 ,$$

an equation in θ_1 which has the (2M+2) solutions

$$\theta_1(r) = \rho \exp\{i\pi r/(M+1)\}, \quad r = 0, 1, \ldots, 2M+1,$$

where $i^2 = -1$ and $\exp\{i\pi r/(M+1)\}$, $r = 0, 1, \ldots, 2M+1$ are the (2M+2), (2M+2)-th roots of unity. Using $\theta_1\theta_2 = \rho^2$, one immediately has

$$\theta_2(r) = \rho \exp\{-i\pi r/(M+1)\} , \quad r = 0, 1, \ldots, 2M+1 .$$

We disregard the case corresponding to r=0 for then $\theta_1(0) = \theta_2(0)$ in violation of the second condition in (17). The relation $\theta_1 + \theta_2 = (\lambda+\mu+\alpha)/\lambda$ implies that for $r = 1, \ldots, 2M+1$

$$\alpha(r) = -(\lambda+\mu) + 2(\lambda\mu)^{1/2} \cos\{\pi r/(M+1)\} . \tag{18}$$

However, notice that $\alpha(r) = \alpha(2M+2-r)$. It thus follows that the non-zero characteristic roots of Q are in the set $\{\alpha(1), \ldots, \alpha(M)\}$.

Replace α and x_j by $\alpha(r)$ and $x_j(r)$ respectively in equation (10) to obtain the following non-trivial solution for each $r = 1, \ldots, M$.

$$x_j(r) = \beta_r[\rho^{j-1} \sin\{\pi r(j+1)/(M+1)\} - \rho^j \sin\{\pi rj/(M+1)\}] \tag{19}$$

where $j = 0, 1, \ldots, M$ and β_r is a constant. It follows that $\alpha(0) = 0$, $\alpha(1), \ldots, \alpha(M)$ are all the (M+1) distinct characteristic roots of Q and that

$$\xi_0 = E_{(M+1)1} , \quad \xi_r = (x_0(r), \ldots, x_M(r))^T$$

where $x_j(r)$ is specified by (19), are the respective right characteristic vectors of Q.

In order to determine the left characteristic vector $\eta_r = (y_0(r), \ldots, y_M(r))^T$ corresponding to the root $\alpha(r)$, $r = 0, 1, \ldots, M$, we have to solve the equations

$$\eta_r^T Q = \alpha(r) \eta_r^T$$

or equivalently the difference equation

$$\lambda y_{j-1} - \{\lambda+\mu+\alpha(r)\}y_j + \mu y_{j+1} = 0, \quad j = 0, 1, \ldots, M, \tag{20}$$

subject to the boundary conditions

$$\lambda y_{-1} = \mu y_0, \quad \lambda y_M = \mu y_{M+1} . \tag{21}$$

Observe that the equation (20) is the same as equation (10), except for an interchange of λ and μ. Thus after some algebra, one has

$$y_j(0) = \gamma_0 \delta^{2j} \tag{22}$$

$$y_j(r) = \gamma_r[\delta^{j+1} \sin\{\pi r(j+1)/(M+1)\} - \delta^j \sin\{\pi rj/(M+1)\}] \tag{23}$$

for $r, j = 0, 1, \ldots, M$, $\delta = \surd(\lambda/\mu)$, γ_r being constants to be so determined that

$$\eta_r^T \xi_r = 1, \qquad r = 0, 1, \ldots, M.$$

Some more simple but tedious algebra yields

$$\gamma_0 = (1-\delta^2)/(1-\delta^{2M+2})$$

$$\beta_r\gamma_r = 2/[(M+1)(1+\delta^2) - 2M\delta \cos\{\pi r/(M+1)\}], \; r = 1, \ldots, M . \tag{24}$$

One finally has

$$p_{jk}(t) = \gamma_0\delta^{2k} + \sum_{r=1}^{M} \beta_r \gamma_r \exp(\alpha(r)t)) \, x_j(r) y_k(r) , \tag{25}$$

which is completely specified by relations (19), (23) and (24).

It is interesting to note that

$$p_k = \lim_{t\to\infty} p_{jk}(t) = \gamma_0 \delta^{2k} , \quad k = 0,1,\ldots, M$$

is independent of j and is in fact the unique solution of the equations

$$(\pi_0, \ldots, \pi_M) P(t) = (\pi_0, \pi_1, \ldots, \pi_M)$$

subject to $\sum_{j=0}^{M} \pi_j = 1$. We shall have more to say about this phenomenon in section 3.6.

Example 2 : Random walk with absorbing barriers.

The random walk with absorbing barriers differs from that with reflecting barriers only in respect of the states 0 and M which are now absorbing states in the sense that $Pr[X(t+h) = k|X(t) = k] = 1$ for $k = 0$ and M for all $h > 0$. Thus the intensity rates for the random walk $\{X(t), t \in IR^+\}$ on the state-space $\{0, 1, \ldots, M\}$ with absorbing barriers at 0 and M are

$$q_{jk} = \begin{cases} \lambda , & k = j+1, \quad j = 1,\ldots, M-1 \\ \mu , & k = j-1, \quad j = 1,\ldots, M-1 \\ 0 , & k \neq j , \quad j = 0, 1, \ldots, M \end{cases}$$

so that $q_{jj} = -(\lambda+\mu)$, $j = 1, \ldots, M-1$ and $q_{00} = q_{MM} = 0$.

The (M+1)-square intensity matrix Q can be expressed in the following partitioned form :

$$Q = \begin{pmatrix} 0 & 0_{M-1} & 0 \\ \xi & R & \eta \\ 0 & 0_{M-1} & 0 \end{pmatrix},$$

where 0_{M-1} is the null row vector of order $(M-1)$, ξ is the $(M-1)\times 1$ column vector $(\mu, 0,\ldots,0)^T$, η is the $(M-1)\times 1$ column vector $(0,\ldots,0,\lambda)^T$ and R is a $(M-1)$-square matrix with $-(\lambda+\mu)$ along the main diagonal, λ along the first super-diagonal, μ along the first sub-diagonal and zero everywhere else. Observe that for all $n \geq 1$,

$$Q^n = \begin{pmatrix} 0 & 0_{M-1} & 0 \\ R^{n-1}\xi & R^n & R^{n-1}\eta \\ 0 & 0_{M-1} & 0 \end{pmatrix},$$

so that in order to obtain Q^n, it is enough to obtain R^n.

The method of obtaining the spectral resolution of R is the same as that adopted in example 1. We only state the results.

The characteristic roots of R are

$$\alpha(r) = -(\lambda+\mu) + 2\surd(\lambda\mu)\cos(\pi r/M), \quad r = 1,\ldots,M-1. \qquad (26)$$

The right and left characteristic vectors,

$$X_r = (x_1(r), \ldots, x_{M-1}(r))^T \quad \text{and} \quad Y_r = (y_1(r), \ldots, y_{M-1}(r))^T$$

corresponding to the root $\alpha(r)$ are specified by

$$x_j(r) = \beta_r \rho^j \sin(\pi jr/M),$$

$$y_j(r) = \gamma_r \delta^j \sin(\pi jr/M), \qquad r, j = 1, \ldots, M-1, \tag{27}$$

where $\rho = \sqrt{(\mu/\lambda)}$, $\delta = \sqrt{(\lambda/\mu)}$, the constants β_r and γ_r being determined by

$$\beta_r \gamma_r = \sum_{j=1}^{M-1} x_j(r)\, y_j(r) = 2/M, \qquad r = 1, \ldots, M-1.$$

Some additional algebra leads to the following expressions for $p_{jk}(t)$:

$$p_{00}(t) = p_{MM}(t) = 1, \qquad p_{0m}(t) = p_{M0}(t) = 0,$$

$$p_{jk}(t) = \frac{2}{M} (\mu/\lambda)^{(j-k)/2} \sum_{r=1}^{M-1} \exp\{\alpha(r)t\} \sin\left(\frac{\pi jr}{M}\right) \sin\left(\frac{\pi kr}{M}\right), \tag{28}$$

$$1 \leq j, k \leq M-1,$$

$$p_{j0}(t) = \frac{2}{M} \sqrt{(\lambda\mu)}\, (\mu/\lambda)^{j/2} \sum_{r=1}^{M-1} [\alpha(r)]^{-1} [\exp\{\alpha(r)t\} - 1] \sin\left(\frac{\pi jr}{M}\right) \sin\left(\frac{\pi r}{M}\right), \tag{29}$$

$$j = 1, \ldots, M-1,$$

$$p_{jM}(t) = \frac{2}{M} \sqrt{(\lambda\mu)}\, (\lambda/\mu)^{(M-j)/2}$$

$$\sum_{r=1}^{M-1} [\alpha(r)]^{-1}[\exp\{t\alpha(r)\}-1]\sin\left(\frac{\pi jr}{M}\right) \sin\left(\frac{\pi r}{M}\right), \tag{30}$$

$$j = 1, \ldots, M-1 .$$

It is interesting to note that $p_{jk}(t) \to 0$ as $t \to \infty$, for $j, k = 1, \ldots, M-1$. One can easily write down $\lim p_{j0}(t)$ and $\lim p_{jM}(t)$ from (29) and (30), but then it is difficult to simplify the resulting expressions. We, therefore, adopt a different approach, which is probabilistically more interesting.

Since the state zero is an absorbing state, in the sense that $p_{00}(t) \equiv 1$ and $p_{0j}(t) \equiv 0$, $j \neq 0$; the event $[X(t) = 0] \subset [X(t+u) = 0]$ for all $u > 0$. Therefore, $p_{j0}(t) \leq p_{j0}(t+u)$ and thus $\pi_j = \lim_{t\to\infty} p_{j0}(t)$ exists and is in fact the probability of ultimate absorption of the Markov process in state zero. A similar argument shows that $\pi_j^* = \lim_{t\to\infty} p_{jM}(t)$ also exists. In order to evaluate π_j, we allow $t \to \infty$ in the Chapman-Kolmogorov equations

$$p_{j0}(t+h) = \sum_{k=0}^{M-1} p_{jk}(h)\, p_{k0}(t)$$

to obtain, after some easy manipulation, that

$$\pi_j\{1 - p_{jj}(h)\}/h = \{p_{jj-1}(h)/h\}\pi_{j+1} + \{p_{jj-1}(h)/h\}\pi_{j-1} .$$

Allowing $h \to 0$, it follows that π_j, $j = 1,\ldots, M-1$ is the solution of the difference equation

$$\lambda \pi_{j+1} - (\lambda+\mu) \pi_j + \mu \pi_{j-1} = 0 , \quad 1 \leq j \leq M$$

subject to the boundary conditions $\pi_0 = 1$, $\pi_M = 0$. Standard techniques immediately yield

$$\pi_j = \{(\mu/\lambda)^j - (\mu/\lambda)^M\}/\{1 - (\mu/\lambda)^M\} , \quad \lambda \neq \mu$$

$1 \leq j \leq M-1$. Following the same argument, one has

$$\pi_j^* = \{1 - (\mu/\lambda)^j\}/\{1 - (\mu/\lambda)^M\} , \quad \lambda \neq \mu, \ 1 \leq j \leq M-1 .$$

If $\lambda = \mu$, $\pi_j = (M-j)/M$ and $\pi_j^* = j/M$, $j = 1, \ldots, M-1$. Note that $\pi_j + \pi_j^* = 1$, which implies that eventually the Markov process is absorbed in state 0 or M.

The reader should refer to Heathcote and Moyal (1959) for a different approach to the problem of obtaining $p_{jk}(t)$ for the random walks with reflecting and absorbing barriers.

Example 3 : Elephant herds

Large mamals, like elephants, are known to wander around in herds of varying sizes. It can happen that when two herds meet, they amalgamate to form a single herd. A herd may also break up into two herds. Holgate (1957) proposed a finite Markov process as a model to describe the formation of herds. One would expect that if the number of herds is large, so that the average size of a herd is small, then the chance of two herds merging into a single herd would be large. Similarly, if the number of herds is small, the chance of a herd breaking up into two herds, would also be large. These considerations suggest the following assumptions.

Suppose a region has M elephants so that there can be 1,2,...,M herds. If at an epoch t, the number $X(t)$ of herds is j, then during (t, t+h], the probability of two of them meeting and amalgamating is $(j-1)\mu h + o(h)$, and the probability of one of them splitting into two herds is $(M-j)\lambda h + o(h)$. As usual we assume that the probability of more than one amalgamations and/or splits during (t, t+h] is o(h), and that these probabilities do not depend on the past of the process before t. It follows that $\{X(t), t \in \mathbb{R}^+\}$ is a birth-death process on the states {1, 2, ..., M} specified by

$$\lambda_j = (M-j)\lambda \ , \quad \mu_j = (j-1)\mu \ , \qquad j = 1, \ldots, M. \tag{31}$$

We now follow Takacs (1960 , p. 96-98) to obtain $p_{jk}(t)$. Let Q denote the M-square intensity matrix corresponding to (31), α a characteristic root of Q and $\eta = (y_0, \ldots, y_{M-1})^T$ the corresponding left characteristic vector. We seek those values of α which provide a non-trivial solution of the equation

$$\eta^T Q = \alpha\eta^T \ ,$$

or equivalently, of the difference equation

$$\lambda(M-j)\ y_{j-1} - \{\lambda(M-j-1) + j\mu\}y_j + (j+1)\mu\ y_{j+1} = \alpha y_j \tag{32}$$

for $j = 0, 1, \ldots, M-1$, with the boundary conditions

$$y_{-1} = y_M = 0\ . \tag{33}$$

Define $V(z) = \sum_{j=0}^{M-1} z^j y_j$ and use (32) and (33) to see that it satisfies the differential equation

$$(\lambda z + \mu)\ (1-z)\ dV(z)/dz = \{\alpha + (M-1)\lambda(1-z)\}\ V(z)\ ,$$

which has the solution

$$V(z) = c(1-z)^{-\alpha/(\lambda+\mu)}\ (\lambda z+\mu)^{M-1+\alpha/(\lambda+\mu)}\ , \qquad (34)$$

where c is an arbitrary constant. Observe that $V(z)$ is a polynomial of degree $(M-1)$ in z, for if the coefficient y_{M-1} of z^{M-1} is zero, then $y_j \equiv 0$, which is not permissible. It follows that non-trivial solutions of (32) are possible only for the following values of α :

$$\alpha(r) = -r(\lambda+\mu)\ , \qquad r = 0,\ 1,\ \ldots,\ M-1\ ,$$

which must therefore be the M distinct characteristic roots of Q. The left characteristic vector

$$\eta_r = (y_0(r),\ \ldots,\ y_{M-1}(r))^T$$

of Q corresponding to $\alpha(r)$ can be recovered from its generating function $V_r(z)$ obtained by replacing α in (34) by $\alpha(r)$, i.e., from

$$V_r(z) = c(r)\ (1-z)^r\ (\lambda z+\mu)^{M-1-r}, \quad r = 0,1,\ldots,M-1. \qquad (35)$$

In what follows we shall assume $c(r) \equiv 1$ without loss of generality.

Let $\xi_r = (x_0(r),\ \ldots,\ x_{M-1}(r))^T$ denote the right characteristic vector of Q corresponding to $\alpha(r)$. Define the M-square matrices X and Y by

$$X = (\xi_0,\ \ldots,\ \xi_{M-1})$$

$$Y = (\eta_0,\ \ldots,\ \eta_{M-1})$$

and note that $XY^T = I_M$, i.e.,

$$\sum_{j=0}^{M-1} x_r(j)\, y_k(j) = \delta_{rk} , \qquad r, k = 0,1,\ldots,(M-1).$$

Hence

$$z^r = \sum_{k=0}^{M-1} z^k \delta_{rk} = \sum_{j=0}^{M-1} x_r(j) \sum_{k=0}^{M-1} z^k y_k(j)$$

$$= \sum_{j=0}^{M-1} x_r(j)\, (\lambda z+\mu)^{M-1-j}(1-z)^j$$

from (35). Substituting $(1-z)/(\lambda z+\mu) = t$, an easy algebra yields the generating function

$$U_r(t) = \sum_{j=0}^{M-1} x_r(j)\, t^j = (\lambda+\mu)^{-(M-1)}\, (1-\mu t)^r\, (1+\lambda t)^{M-1-r}.$$

Equivalently,

$$U_r(t/\mu) = \mu^{-(M-1-r)}\, (\lambda+\mu)^{-(M+1)}\, V_r(t) ,$$

so that it is enough to determine the expansion of V_r to obtain ξ_r and η_r. Finally we get

$$p_{jk}(t) = \sum_{r=0}^{M-1} \exp\{-(\lambda+\mu)\, rt\}\, x_{j-1}(r)\, y_{k-1}(r)$$

$$= \sum_{r=0}^{M-1} \exp\{ -(\lambda+\mu)\, rt\}\, c_{rj}\, y_r(j-1)\, y_{k-1}(r)$$

where $c_{rj} = \mu^{-(M-r-j)}(\lambda+\mu)^{-(M-1)}$,

$$y_j(r) = \mu^{(M-1-r)} \sum_{\nu=0}^{j^*} (-1)^{\nu} (\lambda/\mu)^{j-\nu} \binom{r}{\nu} \binom{M-1-r}{j-\nu},$$

and $j^* = \min\{j, r, M-1-r\}$.

One can easily verify that, as $t \to \infty$, whatever be the initial distribution $X(t)$ converges in law to a random variable X whose distribution is specified by

$$\Pr[X = k] = \binom{M-1}{k-1} \{\lambda/(\lambda+\mu)\}^{k-1} \{\mu/(\lambda+\mu)\}^{M-k}, \quad k=1,\ldots,M. \tag{36}$$

Example 4 : Electric Welders.

Suppose there are M welders who use electric current intermittantly and who work independently of each other. A welder using the electric supply at epoch t, ceases using it during $(t, t+h]$ with probability $\mu h + o(h)$. A welder not using the electric supply at epoch t calls for it during $(t, t+h]$ with probability $\lambda h + o(h)$. Here λ and μ are positive constants and the above mentioned probabilities do not depend on the past of the process before t. Let $X(t)$ denote the number of welders using the electric supply at epoch t. It is easy to verify that all the above assumptions imply that $\{X(t), t \in \mathbb{R}^+\}$ is a birth-death process on $\{0, 1, \ldots, M\}$ with

$$\lambda_j = (M-j)\lambda, \quad \mu_j = j\mu, \qquad 1 = 0,1,\ldots, M.$$

Let $\{Y_M(t), t \in \mathbb{R}^+\}$ denote the birth-death process in example 3 with the state-space $\{1, 2, \ldots, M\}$. It can be readily seen that the birth-death process $\{Y_{M+1}(t)-1,\ t \in \mathbb{R}^+\}$ has the same birth-death rates as the birth-death process $\{X(t),\ t \in \mathbb{R}^+\}$ of this example. One can thus borrow the results from example 3 to obtain $p_{jk}(t)$ and to claim that

$$\lim_{t\to\infty} p_{jk}(t) = \binom{M}{k} \{\lambda/(\lambda+\mu)\}^k \{\mu/(\lambda+\mu)\}^{M-k},\ k = 0,1,\ldots,\ M. \quad (37)$$

The limiting binomial distribution of $X(t)$ prompts one to designate the birth-death process of this and previous example as binomial processes. It is interesting to note that this example can also be looked upon as the continuous time version of the Ehrenfest model of example 1.2.1.

Example 5 : Prendiville or the logistic process.

Consider a population of independently acting individuals such that if at epoch t, there are k individuals in the population $m < k < M$, then in the interval $(t, t+h]$, independently of the past of the population, a specified individual gives birth to a new individual with probability $\lambda(M/k-1)h + o(h)$ or dies with probability $\mu(1-m/k)h + o(h)$, the probability of neither happening being $1 - \lambda(M/k-1)h - \mu(1-m/k)h + o(h)$ where λ and μ are positive constants and m and M, $m < M$, are positive integers. Let $X(t)$ denote the population size at epoch t.

The above assumptions imply that $\{X(t),\ t \in \mathbb{R}^+\}$ is a birth-death process with birth rate $\lambda_j = (M-j)\lambda$ and death-rate $\mu_j = (j-m)\mu$, $j = m,\ldots,\ M$. Once again we find that if $\{Y_M(t),\ t \in \mathbb{R}^+\}$ denotes the birth-death process of example 3 with state space $\{1,\ \ldots,\ M\}$, then, the birth-death process $\{X(t),\ t \in \mathbb{R}^+\}$ is the same as the birth-death process $\{Y_{M-m+1}(t) + m-1,\ t \in \mathbb{R}^+\}$.

6. THE SAMPLE PATHS OF A FINITE MARKOV PROCESS

Let $\{X(t), t \in \mathbb{R}^+\}$ be a Markov process with a finite set S of real numbers as its state-space. Let $(\Omega, \mathbb{F}, P)$ denote the probability space on which the process $\{X(t), t \in \mathbb{R}^+\}$ is defined. By definition 1.5.1 a sample path of the Markov process is the function $X(.,\omega)$ on R^+ to S, where ω is a fixed point in Ω. In this section we study some analytic properties of the sample functions $X(., \omega)$. This study throws light on the evolution of a finite Markov process as time progresses and is also useful in the associated Monte-Carlo studies and problems of statistical inference.

We have already seen in section 2, that one can always construct a finite Markov process with a specified initial distribution and transition probabilities. In this construction the space of all S-valued functions $x(.)$ on $\mathbb{R}^+$ is defined to be Ω. With this construction we have $X(t, x(.)) = x(t)$ and it follows that for this choice of Ω, the concepts of a sample function and the basic point of the underlying sample space Ω coincide. It is with this identification in mind that we use the phrase 'almost all sample functions/paths' to mean all those sample points ω which belong to the complement of a P-null set.

In the study of the sample paths of a continuous parameter stochastic process like a finite Markov process one immediately runs into a technical difficulty. Thus, for example, the set $\{\omega \mid X(u,\omega) = j, t \leq u \leq t+\alpha\}$ or the set $\{\omega \mid X(.,\omega)$ is continuous on $[a, b]\}$, being an uncountable union of events, may not be an event, or even if it is an event, may not have a unique probability assigned to it by the family of finite dimensional distribution of the process. These difficulties can be avoided by assuming that the process is separable in the following sense.

Definition 1 : Let $\{X(t),\ t \in \mathbb{R}^+\}$ be a real stochastic process on the complete probability space $(\Omega,\ \mathbb{F},\ P)$ and let $\mathbb{B}$ be the Borel field in $\mathbb{R}$. Let $\mathbb{D}$ be a sub-class of Borel sets. The process $\{X(t),\ t \in \mathbb{R}^+\}$ is said to be separable relative to the class $\mathbb{D}$ if there exists a sequence $\{t_n,\ n \in Z^+\}$ of non-negative numbers and a P-null event F such that for any $A \in \mathbb{D}$, and an open interval $I \subset [0,\ \infty)$.

$$\{\omega \mid X(t_j,\ \omega) \in A,\ t_j \in I\} - \{\omega \mid X(t,\ \omega) \in A,\ t \in I\} \subset F. \quad (1)$$

The sequence $\{t_n,\ n \in Z^+\}$ is known as the separating sequence. In all later discussions we shall take $\mathbb{D}$ to be the class of all closed intervals and shall call a process separable if it is separable relative to $\mathbb{D}$. It is obvious that if a process is separable, then the completeness of P implies that the set on the left of (1) is an event and thus so is the set $\{\omega \mid X(t,\ \omega) \in A,\ t \in I\}$. The assumption that the finite Markov process is separable one is not very restrictive in view of the following theorem [cf. Chung (1967), p.146].

Theorem 1 : Let $\{X(t),\ t \in \mathbb{R}^+\}$ be a finite Markov process with regular transition probabilities on a complete probability space $(\Omega,\ \mathbb{F},\ P)$. Then there exists a discrete Markov process $\{Y(t),\ t \in \mathbb{R}^+\}$ on $(\Omega,\ \mathbb{F},\ P)$ such that

(i) $P[X(t) = Y(t)] = 1$ for every $t \in \mathbb{R}^+$,

(ii) it is separable with respect to any sequence dense in $[0,\ \infty)$.

Remark : An immediate consequence of (i) is that the processes $\{X(t), t \in \mathbb{R}^+\}$ and $\{Y(t), t \in \mathbb{R}^+\}$ have the same family of finite dimensional distributions. Such processes are said to be equivalent versions of each other. A seperable process satisfying (ii) is called a well-seperable process. Thus the above theorem asserts that every finite Markov process has a well-seperable, equivalent version. In what follows we shall therefore assume without additional emphasis, that our finite Markov process is well-seperable.

Lemma 1 : Let $\{X(t), t \in \mathbb{R}^+\}$ be a finite Markov process. Then

$$\lim_{t \to u} \Pr[X(u) \neq X(t)] = 0$$

iff the process has regular transition probabilities.

Proof : Let $\{X(t), t \in \mathbb{R}^+\}$ have regular transition probabilities and let $v = \min(t, u)$. Then

$$\Pr[X(u) \neq X(t)] = \sum_{j \in S} \Pr[X(v) = j]\{1 - \Pr[X(t-u) = j]\} \tag{2}$$

$$\leq \max_{j \in S} \{1 - p_{jj}(t-u)\} \to 0$$

as $t \to u$.

The converse implication is an immediate consequence of (2). The proof is complete.

The above lemma can be interpreted to assert that a discrete Markov process is continuous in probability on $\mathbb{R}^+$ iff it has regular transition probabilities. The following theorem shows that, in fact, almost all the sample paths of a finite Markov process are continuous at every $t \in (0, \infty)$.

Theorem 2 : Let $\{X(t), t \in \mathbb{R}^+\}$ be a finite Markov process on a probability space $(\Omega, \mathbb{F}, P)$. Then

(i) for every state $j \in S$ and epoch $t \in \mathbb{R}^+$

$$\Pr[X(u) = j, t \leq u \leq t+\alpha \mid X(t) = j] = \exp[-q_j\alpha], \quad \alpha > 0,$$

and (ii) almost all the sample paths are continuous at every $t \in (0, \infty)$.

Proof : Consider the sequence of all real numbers of the type $k\alpha/2$, $k = 0, 1, \ldots, 2^n$, $n = 1, 2, \ldots$ using which as the seperating sequence for our well-seperable finite Markov process, we have

$$\Pr[X(u) = j, \; t \leq u \leq t+\alpha \mid X(t) = j]$$

$$= \lim_{n\to\infty} \Pr[X(t + k\alpha/2^n) = j, k = 0,1,\ldots,2^n \mid X(t)=j]$$

$$= \lim_{n\to\infty} \{p_{jj}(\alpha/2^n)\}^{2^n} = \exp[-q_j\alpha],$$

where the last step is a consequence of remark 3 following theorem 2.1. This establishes (i).

In order to establish (ii), let t_0 be a fixed point in $\mathbb{R}^+$ and let $X(t_0) = j$. Observe that

$$[\omega \mid \lim_{t\to t_0} X(t,\omega) = j] = \bigcap_{\varepsilon>0} \; \bigcup_{\delta>0} \; \bigcap_{|t-t_0|<\delta} [|X(t) - j| < \varepsilon]$$

$$= \bigcup_{\delta>0} \; \bigcap_{|t-t_0|<\delta} [X(t) = j]$$

$$= \bigcup_{n\geq 1} \; \bigcap_{|t-t_0|\leq 1/n} [X(t) = j],$$

where (3) is a consequence of the discrete nature of the r.v.s X(t). Since the sequence

$$\bigcap_{|t-t_0|\le 1/n} [X(t) = j],\ n = 1, 2, \ldots$$

is a non-decreasing sequence of events

$$\Pr[\lim_{t\to t_0} X(t) = j \mid X(t_0) = j]$$

$$= \lim_{n\to\infty} \Pr[X(t) = j,\ t_0 - 1/n \le t \le t + 1/n \mid X(t_0) = j]$$

$$= \lim_{n\to\infty} \Pr[X(t_0 - 1/n) = j] \exp[-2q_j/n] / \Pr[X(t_0) = j]$$

$$= 1$$

by virtue of the continuity of the transition probabilities. The proof is complete.

The result (ii) of the above theorem does not **guarantee** that almost every sample function of the discrete Markov process is continuous on any non-degenerate interval [a, b] of IR^+. In order to understand more clearly the nature of the sample functions we need the following definitions.

Definition 2 : A real-valued function g(.) on IR^+ is a step function if

(i) it has finitely many points of discontinuities in every finite closed interval $[a, b] \subset IR^+$,

(ii) it is identically equal to a constant in every open

interval $(a, b) \subset \mathbb{R}^+$ if the end points a and b are points of continuity of $g(.)$, and

(iii) for every point t_0 of discontinuity of g, the limits $g(t_0 + 0)$ exist and either

$$g(t_0 \pm 0) \leq g(t_0) \leq g(t_0 + 0) \quad \text{or} \quad g(t_o - 0) \geq g(t_0) \geq g(t_0+0) \tag{4}$$

with at least one strict inequality.

Definition 3 : A real valued function $g(.)$ on $\mathbb{R}^+$ is said to have a jump at a point $t \in \mathbb{R}^+$ if the limits $g(t_0 - 0)$ and $g(t_0 + 0)$ exist and satisfy one of the inequalities in (4), with at least one strict inequality.

It is obvious from the above definitions that all the points of discontinuities of a step function which are in the interior of $\mathbb{R}^+$ are jumps. It is also obvious that a function $g(.)$ with a finite range space and continuous except for jumps is a step functior

Theorem 3 : Let $\{X(t), t \in \mathbb{R}^+\}$ be a finite Markov process.

(i) If $q_j = 0$ and $X(t_0) = j$, then

$$P[X(t) = j, \; t > t_0 \mid X(t_0) = j] = 1 .$$

(ii) If $q_j > 0$ and $X(t_0) = j$, then with probability one, there exists a sample function discontinuity for some $t > t_0$ and this discontinuity is a jump.

(iii) Let $0 < \alpha < \infty$, $X(t_0) = j$ and suppose that there exists at least one discontinuity in the sample path in the interval $[t_0, t_0 + \alpha)$. Then the conditional probability that the first discontinuity is a jump to k, given the information stated earlier, is q_{jk}/q_j, q_j being assumed to be positive.

Proof : The statement (i) is a trivial consequence of statment (i) of theorem 2. In the rest of the proof, we therefore assume that $q_j > 0$, and without loss of generality take $t_0 = 0$.

In order to prove (ii), let $n \geq 2$, $\alpha, \beta > 0$, $k \in S - \{j\}$, and define the event

$$\Lambda_{n\beta}(k) = \sum_{r=2}^{n} [X(u) = j,\ 0 \leq u \leq (r-1)\alpha/n,\ X(v) = k,\ \frac{r\alpha}{n} \leq v \leq \frac{r\alpha}{n} + \beta].$$

Then by (i) of theorem 2 and the Markov property

$$P[\Lambda_{n\beta}(k) \mid X(0) = j]$$

$$= \sum_{r=2}^{n} \exp[-(r-1)\alpha q_j/n]\ p_{jk}(\alpha/n) \exp(-q_k\beta)$$

$$= p_{jk}(\alpha/n) \exp(-q_k\beta) \{\exp(-\alpha q_j/n) - \exp(-\alpha q_j)\}$$

$$\times \{1 - \exp(-\alpha q_j/n)\}^{-1}$$

$$\rightarrow \exp(-q_k\beta) \{1 - \exp(-\alpha q_j)\}\ q_{jk}/q_j \tag{5}$$

as $n \rightarrow \infty$, by virtue of remark 3 following theorem 2.1.

Let

$$\Lambda_\beta(k) = \limsup_{n \to \infty} \Lambda_{n\beta}(k) = \bigcap_{n=1} \bigcup_{r \geq n} \Lambda_{r\beta}(k)$$

denote the event of ω points which belong to $\Lambda_{n\beta}(k)$ for infinitely many values of n. One can easily see that

$$\Pr[\limsup \Lambda_{n\beta}(k) \mid X(0) = j]$$

$$\geq \limsup \Pr[\Lambda_{n\beta}(k) \mid X(0) = j]$$

$$= \lim \Pr[\Lambda_{n\beta}(k) \mid X(0) = j]$$

$$= \{1 - \exp(-q_j\alpha)\} \exp(-q_j\beta) q_{jk}/q_j \tag{6}$$

by virtue of (5) and Fatou's lemma [cf. Loève (1968), p. 125].

Since $\Lambda_\beta(k) \supset \Lambda_\gamma(k)$, when $\beta < \gamma$, it also follows that if $\Lambda(k)$ denotes the event $\lim \Lambda_\beta(k)$ obtained as $\beta \downarrow 0$, then

$$\Pr[\Lambda(k) \mid X(0) = j] = \lim_{\beta \downarrow 0} \Pr[\Lambda_\beta(k) \mid X(0) = j]$$

$$\geq \{1 - \exp(-q_j\alpha)\} q_{jk}/q_k.$$

Let $\omega \in \Lambda(k)$. Then there exists a number $\tau_1(\omega) \in (0,\alpha)$ such that $X(t,\omega) = j$, $0 \leq t < \tau_1(\omega)$ and $X(t,\omega) = k$, $\tau_1(\omega) < t < \alpha$. Hence

$$\bigcup_{k \in S-\{j\}} \Lambda(k) = [X(u) = j, \quad 0 \leq u \leq \alpha)]^c$$

and thus equality must hold in the following inequality :

$$\sum_{k\varepsilon S-\{j\}} \Pr[\Lambda(k) \mid X(0) = j\} \geq \{1 - \exp(-q_j\alpha)\}q_{jk}/q_j$$

$$= 1 - \Pr[X(u) = j,\ 0 \leq u \leq \alpha]. \qquad (7)$$

We must thus also have

$$\Pr[\Lambda(k) \mid X(0) = j] = [1 - \exp(-q_j\alpha)]\ q_{jk}/q_j. \qquad (8)$$

One immediate implication of equality in (7) is that the probability of a sample path discontinuity in the interval $(0,\alpha)$, conditional on $X(0) = j$ is $1 - \exp(-q_j\alpha)$ which tends to one as $\alpha \to \infty$. This establishes that if $q_j > 0$ and $X(0) = j$, then almost every sample path $\{X(t,\omega),\ 0 < t < \infty\}$ has at least one discontinuity. In order to show that this discontinuity is a jump in the sense of definition 3, one has to recall that a consequence [cf. Doob (1953), p. 50-71] of the assumption of seperability of the stochastic process $\{X(t),\ t \varepsilon \mathbb{R}^+\}$ is that for any open interval $I \subset \mathbb{R}^+$, there exists a denumerable set $H \subset I$, such that

$$\Pr[\sup_{t\varepsilon I} X(t) = \sup_{t\varepsilon H} X(t)] = 1 = \Pr[\inf_{t\varepsilon I} X(t) = \inf_{t\varepsilon H} X(t)]\ .$$

It follows that in view of the almost sure continuity of $X(t)$ on the denumerable set H,

$$X(\tau_1(\omega),\omega)\ \varepsilon\ [X(\tau_1(\omega) - 0,\omega),\ \ X(\tau_1(\omega) + 0,\omega)]$$

holds for almost all ω. This proves that the discontinuity of $X(t)$ at τ_1 is almost surely a jump.

The last part (iii) of the theorem is an immediate consequence of (8) and (i) of Theorem 2.

We are now in a position to establish the following theorem which completes the description of the sample paths of a finite Markov process. It is necessary here to emphasize that our finite Markov process by assumption is a well seperable stochastic process with standard stationary probabilities.

Theorem 4 : The sample functions of a finite Markov process are almost all step functions.

Proof : We have seen in theorem 3, that if $X(0) = j$ and $q_j = 0$, then $X(t) = j$ for all $t > 0$ with probability one. Hence let $q_j > 0$ in which case there is almost surely a first discontinuity which is a jump at some random epoch τ_1. If $X(\tau_1+0) = k$ and $q_k=0$, it follows by the strong Markov property that

$$Pr[X(t) = k \mid X(\tau_1+0) = k] = 1 \quad \text{for all} \quad t > \tau_1 .$$

In case $q_k > 0$, one can invoke theorem 3 to gurantee with probability one, that there is a jump discontinuity at some random epoch $\tau_2 > \tau_1$. Thus unless the Markov process visits a state k, with $q_k = 0$, we shall get a non-terminating increasing sequence $\{\tau_n, n \geq 1\}$ of random epochs. Let us therefore agree that if for some epoch τ_n, $q_{X(\tau_n+0)} = 0$, then we put $\tau_{n+1} = \tau_{n+2} = \dots = \infty$. In this situation, it is obvious that the theorem is true.

It is therefore enough to consider the case obtaining when $q_j > 0$ for all $j \in S$. The argument of theorem 3 then garantees that

$$\Pr[\tau_{n+1} - \tau_n \geq \alpha \mid X(\tau_n + 0) = j] = \exp(-q_j\alpha)$$

for all $\alpha \geq 0$. In particular, if $q = \max\{q_j, j \in S\}$, then for every $\alpha \geq 0$,

$$\Pr[\tau_{n+1} - \tau_n \geq \alpha]$$

$$= \sum_{k \in S} \Pr[X(\tau_n+0) = k] \Pr[\tau_{n+1} - \tau_n \geq \alpha \mid X(\tau_n+0) = k]$$

$$= \sum \exp(-q_k\alpha) \Pr[X(\tau_n + 0) = k]$$

$$\geq \exp(-q\alpha) .$$

It follows by Fatou's lemma [cf. Loève (1968), p. 125] that

$$\Pr[\limsup_{n \to \infty} \tau_{n+1} - \tau_n \geq \alpha] \geq \limsup \Pr[\tau_{n+1} - \tau_n \geq \alpha]$$

$$\geq \exp(-q\alpha) .$$

This inequality is easily seen to imply that

$$\Pr[\lim \tau_n = \infty] \geq \exp(-q\alpha)$$

for every $\alpha \geq 0$. In particular it follows that

$$\Pr[\lim \tau_n = +\infty] = 1 .$$

The theorem is established.

The results of this section can be used to obtain a realization $\{X(u), 0 \leq u \leq t\}$ of the Markov process with initial distribution $\{p_i(0),\ i \in S\}$, intensity rates $q_{ij},\ i \neq j,\ i,j \in S$ and state-space $S = \{1, \ldots, M\}$. The procedure for obtaining a realization is described by the following steps :

Step 1 : Obtain a random observation from the initial distribution $\{p_1(0), \ldots, p_M(0)\}$. Suppose that $X(0) = i_0$.

Step 2 : If $q_{i_0} = 0$ i.e., if state i_0 is absorbing, stop the procedure and then $\{X(u) = i_0,\ 0 \leq u \leq t\}$ is the realization. If $q_{i_0} > 0$, proceed to step 3.

Step 3 : Draw a random observation from the exponential distribution with density function

$$f(x) = q_{i_0} \exp(-q_{i_0} x) , \quad 0 < x < \infty .$$

Let this observation be t_0. If $t_0 > t$, stop the procedure. The realization is $\{X(u) = i_0,\ 0 \leq u \leq t\}$. If $t_0 < t$, proceed to step 4.

Step 4 : If $t_0 < t$ and $q_{i_0} > 0$, obtain a random observation from the discrete distribution $\{r_{i_0 1}, \ldots, r_{i_0 M}\}$, where $r_{jk} = q_{jk}/q_j$ if $q_j > 0$. Suppose this observation is i_1.

Step 5 : If $q_{i_1} = 0$, stop the procedure and the required realization is $\{X(u) = i_0,\ 0 \leq u \leq t_0,\ X(u) = i_1,\ t_0 < u \leq t\}$. If $q_{i_1} > 0$, repeat step 3 with i_0 replaced by i_1. The subsequent steps are the appropriate modifications of step 4.

CHAPTER III

CLASSIFICATION OF STATES

1. INTRODUCTION

We have given a brief description of the classification of states of a Markov chain in section 1.2. In this chapter we discuss the problem of classifying the states of a discrete Markov process $\{X(t), 0 \quad t \in \mathbb{R}^+\}$. The concepts related to the classification of states of a Markov chain are used to classify the states of a discrete Markov process also. In fact, we find that it is possible to classify the states of a discrete Markov process by classifying the states of some related Markov chains introduced earlier. We study some of their properties in section 2. In section 3, we look at the properties of accessibility of the states of and irreducibility of a discrete Markov process. In section 4, criteria for classifying states as persistent or transient are discussed. Section 5 deals with the asymptotic properties of the transition probabilities. We discuss invariant distribution of a finite Markov process in section 6 and give some examples in section 7.

2. ASSOCIATED MARKOV CHAINS

Let $\mathbf{X} = \{X(t), t \in R^+\}$ be a finite Markov process with state-space S and standard transition probabilities $p_{jk}(t)$, $t \in \mathbb{R}^+$, $j, k \in S$. We associate the following three Markov chains with the process **X** :

(i) The discrete skeleton $\mathbf{X}(h) = \{X(nh), n \in Z^+\}$, $h > 0$, to scale h defined on the state-space S and with one-step transition probabilities $p_{jk}(h)$, [cf. Definition 2.2.2];

(ii) the Markov chain $\mathbf{X}(P) = \{Y_n, n \in Z^+\}$ with state space S and one-step transition probabilities

$$p_{jk} = \delta_{jk} + q_{jk}/\beta \quad , \qquad j, k \in S , \tag{1}$$

where $\max\{q_j \mid j \in S\} < \beta < \infty$. [cf. Theorem 2.3.2]; and δ_{jk} is the usual Kronecker delta;

(iii) the Markov chain $\mathbf{X}(R) = \{Z_n, n \in Z^+\}$ with state-space S and one-step transition probabilities

$$r_{jk} = \begin{cases} (1 - \delta_{jk})\, q_{jk}/q_j, & \text{if } q_j > 0, \\ \delta_{jk} \quad , & q_j = 0 . \end{cases} \tag{2}$$

The matrix $R = ((r_{jk}))$ is known as the jump matrix [cf. Remark 2.3.4].

Although $\mathbf{X}(h)$ is defined on the same probability space as that of **X**, it is not necessary that $\mathbf{X}(P)$ and $\mathbf{X}(R)$ be also defined on the probability space of **X**. The connections between **X** and $\mathbf{X}(h)$ and **X** and $\mathbf{X}(P)$ have already been discussed in section 2.3 and Theorem 2.3.2 respectively. In this section we proceed to bring out the relation between **X** and $\mathbf{X}(R)$.

Lemma 1 : Let $\{\tau_n, n \in Z^+\}$ be the sequence of random epochs of jumps of the **X**-process, with $\tau_0 = 0$. Then $\{X(\tau_n + 0), n \in Z^+\}$ is a Markov chain on S with one-step transition probabilities r_{jk}, $j, k \in S$ defined by (2).

Proof : This is an immediate consequence of the strong Markov property of **X** and Theorem 2.6.3.

It should be noted that if for some $n \in Z^+$, $X(\tau_n + 0) = j$ and $q_j = 0$, then we take $\tau_{n+1} = \tau_{n+2} = \dots = \infty$ and write $X(\tau_{n+1}) = X(\tau_{n+2}) = \dots = j$. In view of this lemma, the finite Markov chain $\mathbf{X}(R)$ is essentially the process **X** observed immediately after the epochs of jumps only. It is for this reason that the matrix $R = ((r_{jk}))$ is sometimes refered to as the jump matrix of the process **X**. However, it is possible to connect the n-step transition probabilities $r_{jk}^{(n)}$, $n \in Z^+$, with the transitions of the **X**-process in a more meaningful manner.

Let $N(s, s+t)$ be the number of jumps of the process **X** during the interval $(s, s+t]$ and let for $j, k \in S$, $n \in Z^+$,

$$p_{jk}(t, n) = Pr[X(s+t) = k, \ N(s, s+t) = n \mid X(s) = j] .$$

A method of calculating the probabilities $p_{jk}(t, n)$ in a recursive fashion is provided by the following

Lemma 2 : We have

$$p_{jk}(t, 0) = \delta_{jk} \exp(-q_j t) \tag{3}$$

and for $n \geq 0$,

$$p_{jk}(t, n+1) = \sum_{\nu \varepsilon S-\{j\}} \int_0^t q_{j\nu} p_{\nu k}(t-u, n) \exp(-q_j u) du. \tag{4}$$

Proof : The expression (3) for $p_{jk}(t, 0)$ is an immediate consequence of its definition and (i) of Theorem 2.6.2. In order to obtain (4), observe that if $X(s) = j$

$$\rho_j = \inf \{t \mid t > 0, X(s+t) \neq j, X(s) = j\}$$

denoting the so-called exit time from state j is an r.v. as a consequence of the assumption of the seperability of the **X**-process. Moreover, if $q_j > 0$, by theorem 2.6.2, ρ_j has exponential distribution with mean $1/q_j$. We thus have by the theorem of total probabilities and the strong Markov property that

$$p_{jk}(t, n+1) = \sum_{\nu \varepsilon S-\{j\}} \int_0^t \Pr[X(s+t) = k, N(s+u, s+t) = n | X(s+u+0) = \nu]$$

$$\times \ (q_{j\nu}/q_j) \ dP[\rho_j \leq u]$$

$$= \sum_{\nu \varepsilon S-\{j\}} \int_0^t q_{j\nu} \exp(-q_j u) \, p_{\nu k}(t-u, n) \, du,$$

which is (4).

Lemma 3 : For all $j, k \varepsilon S$

$$p_{jk}(t) = \sum_{n=0}^{\infty} p_{jk}(t, n) . \tag{5}$$

Proof : This is an immediate consequence of the theorem of total probabilities.

The following theorem provides a connection between r_{jk} and $p_{jk}(t, n)$.

Theorem 1 : For every $n \in Z^+$,

$$r_{jk}^{(n)} = \int_0^\infty q_k\, p_{jk}(t, n)\, dt\,. \tag{6}$$

Proof : This result is easily seen to be true for $n = 0$ by using (3). Assume that it holds for some $n > 0$. Then by (4) we have

$$\int_0^\infty p_{jk}(t, n+1)\, q_k dt$$

$$= \sum_{\nu \in S-\{j\}} \int_0^\infty q_k \{ \int_0^t q_{j\nu} p_{\nu k}(t-u, n) \exp(-q_j u) du\} dt$$

$$= \sum_{\nu \in S-\{j\}} \int_0^\infty q_k\, q_{j\nu} \exp(-q_j u) \{\int_u^\infty p_{\nu k}(t-u, n)\, dt\}\, du$$

$$= \sum_{\nu \in S-\{j\}} \int_0^\infty q_{j\nu} \exp(-q_j u) \{ \int_0^\infty q_k\, p_{\nu k}(t, n)\, dt \} du$$

$$= \sum_{\nu \in S} r_{j\nu}\, r_{\nu k}^{(n)} = r_{jk}^{(n+1)}\,,$$

as required.

3. ESSENTIAL AND INESSENTIAL STATES

Let $\{X(t), t \in \mathbb{R}^+\}$ be a discrete Markov process with state-space S and transition probabilities $p_{jk}(t)$, $j, k \in S$.

Definition 1 : A state j leads to a state k, $j \to k$, or k is accessible from j, if for some $t > 0$, $p_{jk}(t) > 0$. Two states j and k communicate with each other ($j \leftrightarrow k$) if $j \rightarrow k$ and $k \rightarrow j$.

It is easy to verify that the relation "$\rightarrow$" is reflexive ($j \rightarrow j$) and transitive ($j \rightarrow k$, $k \rightarrow \ell$ imply $j \rightarrow \ell$). However, in general it is not symmetric, e.g. in the case of the random walk with absorbing barriers, discussed in example 2.5.2, every state j, $0 < j < M$, leads to state zero, but zero does not lead to any j. On the other hand the communication relation $\leftrightarrow$ is an equivalence relation i.e., it is reflexive, symmetric and transitive. This equivalence relation partitions the state-space into equivalence classes. Two states belong to the same equivalence class iff they communicate. In what follows we shall refer to an equivalence class as communicating class or simply a class. The class containing a specified state j will be denoted by $C(j)$ and is the class of all states $k \in S$ which communicate with j.

Example 1 : In example 2.5.2, describing the random walk with absorbing barriers, the communicating classes are $\{0\}, \{1, \ldots, M-1\}$ and $\{M\}$. In the random walk with reflecting barriers [cf. example 2.5.1] there is only one class viz. $\{0, \ldots, M\}$.

Definition 2 : A discrete Markov process is irreducible if the entire state-space is a single communicating class. If a process is not irreducible, it is said to be reducible.

Example 2 : The random walk with reflecting barriers is an irreducible process while that with absorbing barriers is a reducible process.

In general, it is easier to discuss the irreducibility or otherwise of a Markov chain than that of a process. We, therefore, discuss the accessibility properties of states in the four processes

X, **X**(h), **X**(P) and **X**(R) and their inter-relations in the following

Theorem 1 : If $j \to k$ in any one of the processes **X**, **X**(h), **X**(P) and **X**(R), and S is finite, then $j \to k$ in all the other processes.

Proof : Suppose $j \to k$ in **X**. Then by lemma 2.2.2 $p_{jk}(t) > 0$ for all $t > 0$. In particular, $p_{jk}(h) > 0$ and therefore $j \to k$ in **X**(h).

If $j \to k$ in **X**(h), the relation [cf. equation (2.3.6)]

$$0 < p_{jk}(h) = \sum_{n=0}^{\infty} \exp(-\beta h)\ (\beta h)^n\ p_{jk}^{(n)}/n\ !$$

implies that $p_{jk}^{(n)} > 0$ for at least one n and thus $j \to k$ in **X**(h) implies $j \to k$ in **X**(P).

Suppose now that $p_{jk}^{(n)} > 0$ for some $n > 0$. Then there exist distinct states $i_0, i_1, \ldots, i_m$, $m < M$, such that $i_0 = j$, $i_m = k$ and $\prod_{k=1}^{m} p_{i_{k-1}i_k} = \Pi\{q_{i_{k-1}i_k}/\beta\} > 0$ which in turn implies that $q_{i_0}, \ldots, q_{i_{m-1}}$ are all positive. Thus $r_{jk}^{(m)} \geq \Pi\{q_{i_{k-1}i_k}/q_{i_{k-1}}\} > 0$ and we conclude that if $j \to k$ in **X**(P), then $j \to k$ in **X**(R).

Finally if $r_{jk}^{(n)} > 0$ for some $n > 0$, then $p_{jk}(t, n) > 0$ by equation (2.6). The relation (2.5) thus yields the fact that if $j \to k$ in **X**(R) then $j \to k$ in **X**. The proof is complete.

Definition 3 : If $C(j) = \{j\}$, i.e. if $j \not\to k$ for any $k \neq j$, then state j is absorbing or equivalently state j is absorbing iff $p_{jj}(t) = 1$ for all $t \in \mathbb{R}^+$.

Corollary : If a state j is absorbing in one of the four processes

$\mathbf{X}$, $\mathbf{X}(h)$, $\mathbf{X}(P)$, $\mathbf{X}(R)$, then it is absorbing in all other processes. In particular, j is absorbing iff $q_j = 0$.

Definition 4 : A state j of a discrete Markov process is essential if it communicates with every state to which it leads. A state which is not essential is called an inessential state.

Example 3 : Every state in the random walk with reflecting barriers is essential. All states j, $1 \le j \le M-1$ in the random walk with absorbing barriers are inessential.

We now establish a solidarity theorem [cf. Section 1.2] .

Theorem 2 : Being essential is a solidarity property.

Proof : Let j be an essential state and let $j \to k$. If $k \to \ell$ then $j \to k \to \ell$ implies that $j \to \ell$ and since j is essential, $\ell \to j \to k$. Thus $\ell \to k$ and therefore k is essential.

Corollary : Being inessential is a solidarity property.

Remark : A state j is inessential iff there exists a state k such that $j \to k$ but $k \not\to j$. The above theorem and its corollary imply that essential (inessential) states communicate only with essential (inessential) states. However, the example of the random walk with absorbing barriers shows that an inessential state may lead to an essential state.

Definition 5 : A proper subset C of the state-space S is said to be closed if for every $j \in C$ and $k \in S - C$, $j \not\to k$. The state-space S is closed by definition. A closed set C is minimal closed if no proper subset of C is closed.

Observe that the properties of being essential or inessential or the closure of a set depend only on accessibility of states. Hence in view of theorem 1 if these properties hold in any one of the four Markov processes, they hold in all the other three. In particular, if a state j is absorbing in one of the processes, it is absorbing in all the other processes.

<u>Theorem 3</u> : Let j be an essential state and let

$$C(j) = \{k \mid k \in S, \ j \to k\}$$

be the class of all states k which are accessible from j. Then $C(j)$ is minimal closed.

<u>Proof</u> : It is obvious that by the definition of an essential state, any two states in $C(j)$ communicate and therefore $C(j)$ is a class. If $k \in C(j)$ and $\ell \notin C(j)$, then $k \not\to \ell$ for if $k \to \ell$, $j \to \ell$ and then $\ell \in C(j)$, a contradiction. Hence $C(j)$ is a closed essential class.

Suppose, if possible, a proper subset A of $C(j)$ is closed. Then for every $k \in A$, and every $\ell \in C(j) - A$, $k \not\to \ell$, which contradicts the fact that $C(j)$ is a class. Hence $C(j)$ is minimal closed and the proof is complete.

By definition an inessential class can not be closed. The following theorem demonstrates that any finite union of inessential classes is <u>not</u> closed. However, a countable union of inessential classes may be closed.

<u>Theorem 4</u> : A finite union of inessential classes is not closed.

<u>Proof</u> : Let $A_1, \ldots, A_N$ be $N < \infty$ inessential classes. Let

$j_1 \in A_1$. There exists a state $J_2 \notin A_1$ such that $j_1 \to j_2$, but $j_2 \not\to j_1$. If $j_2 \notin A_2 \cup \cdot \cup A_N$, the theorem is established. So let $j_2 \in A_2 \cup \ldots \cup A_n$ and in particular, without loss of generality let $j_2 \in A_2$. Then there exists a state j_3, such that $j_2 \to j_3$ but $j_3 \not\to j_2$. Then $j_2 \notin A_1$ for if $j_3 \in A_1$, $j_1 \to j_2 \to j_3 \to j_1$ implying $j_2 \to j_1$ which is contradiction. So if $j_3 \notin A_3 \cup \ldots \cup A_N$ our purpose is served and if $j_3 \in A_3 \cup \ldots \cup A_N$ we have to repeat the above argument. In any case, since $N < \infty$, we must be able to find states $j \in \cup A_r$ and $k \notin \cup A_r$, such that $j \to k$ and $k \not\to j$. Thus $\bigcup_{r=1}^{N} A_r$ is not closed.

<u>Corollary</u> : All states of a finite Markov process can not be inessential or equivalently every finite Markov process has at least one essential state.

The above theorem 2 and its corollary imply that for every finite Markov process there must be at least one essential class. Hence, in general we can partition the finite state-space S containing M states of a Markov process into a finite number of classes, $C_1, \ldots, C_r$ and $D_1, \ldots, D_s$, such that $r + s \leq M$, $1 \leq r \leq M$, $0 \leq s < M$, where $C_1, \ldots, C_r$ are minimal closed essential classes and $D_1, \ldots, D_s$ are inessential classes. If $j \in \bigcup_{\nu=1}^{s} D_\nu$, there exists a state $k \in \bigcup_{\nu=1}^{r} C_\nu$, such that $j \to k$, but no state in C_ν, $1 \leq \nu \leq r$, leads to any state outside C_ν. By a suitable renumbering of states, if necessary, we can represent the matrix P(t) (and therefore matrices P(h), P and R) in the following partitioned form :

$$P(t) = \begin{pmatrix} P_1(t) & 0 \ldots 0 & 0 & . & 0 \\ 0 & P_2(t) \ldots 0 & 0 & . & 0 \\ . & . \quad . & . & & . \\ 0 & 0 \ldots P_r(t) & 0 & . & 0 \\ U_{11}(t) & U_{12}(t) \ldots U_{1r}(t) & U_1(t) & . & 0 \\ . & . \quad . & . & & . \\ U_{s1}(t) & U_{s2}(t) \ldots U_{sr}(t) & 0 & . & U_s(t) \end{pmatrix},$$

where $P_1(t), \ldots, P_r(t)$ are stochastic matrices and $U_1(t), \ldots U_s(t)$ are sub-stochastic matrices in the sense that they are square matrices with non-negative elements, whose row sums are atmost equal to 1. This is sometimes referred to as the canonical form of $P(t)$.

We close this section by demonstrating that a discrete Markov process can not have periodic states. In view of lemma 2.2.2, in a finite Markov process, $p_{jj}(t) > 0$ for all $t > 0$ and therefore $p_{jj}(t)$ cannot vanish outside a lattice sub-set of $\mathbb{R}^+$. In fact, one can easily verify that every state of $\mathbf{X}(h)$ and $\mathbf{X}(P)$ is aperiodic. However, this is not true of the states of $\mathbf{X}(R)$. For example in case of the two-state Markov process $\{X(t), t \quad \mathbb{R}^+\}$ with intensity matrix

$$Q = \begin{pmatrix} -\lambda & \lambda \\ \mu & -\mu \end{pmatrix} \qquad 0 < \lambda , \ \mu < \infty ,$$

the jump matrix

$$R = \begin{pmatrix} 0 & 1 \\ 1 & 0 \end{pmatrix}$$

and thus every state of $\mathbf{X}(R)$ is periodic with period 2.

4. PERSISTENT AND TRANSIENT STATES

In section 1.2 we have defined a state j of a Markov chain to be persistent (transient) according as the probability f_{jj} of atleast one return to the state j is one (less than one). One can show that j is persistent (transient) iff the series $\sum_{n=1}^{\infty} p_{jj}^{(n)}$ converges to $+\infty$ (a finite limit). A persistent state j is further classified as non-null (null) if lim sup $p_{jj}^{(n)}$ is positi (zero). In general the sequence $\{ p_{jk}^{(n)}, n \varepsilon Z^{+}\}$ does not have a limit. However, if k is an aperiodic state, $\alpha_{jk} = \lim_{n\to\infty} p_{jk}^{(n)}$ exists. If k is persistent null or transient or if $j \not\to k$, then $\alpha_{jk} = 0$. If $j \to k$ and k is an aperiodic persistent, non-null state i.e., if it is an ergodic state, then α_{jk} is positive and in fact equals f_{jk}/μ_k, where μ_k is the mean recurrence time of state k. We refer the interested reader to Chung (1967), p. 21-34, for proofs of the various statements made above. In this section we extend the above concepts and results to discrete Markov processes. In this exercise we need the following theorem which is of independent interest.

Theorem 1 : Let $\mathbf{X} = \{X(t), t \varepsilon \mathbb{R}^{+}\}$ be a discrete Markov process with state space S and transition probabilities $p_{jk}(t)$, $j,k \varepsilon S$. Then

$$\alpha_{jk} = \lim_{t\to\infty} p_{jk}(t) \tag{1}$$

exists for all $j, k \varepsilon S$. If the state-space is finite,

$$\alpha_{jk} = \sum_{\nu \varepsilon S} \alpha_{j\nu} p_{\nu k}(t) = \sum_{\nu \varepsilon S} p_{j\nu}(t)\, \alpha_{\nu k} = \sum_{\nu \varepsilon S} \alpha_{j\nu}\, \alpha_{\nu k} \tag{2}$$

for all $t \varepsilon \mathbb{R}^{+}$, $j, k \varepsilon S$.

Proof : If $p_{jk}(t) = 0$ for some $t > 0$, then $p_{jk}(t) \equiv 0$ and therefore $\alpha_{jk} = 0$. So let $p_{jk}(t) > 0$ for all $t > 0$.

Consider the discrete skeleton $\mathbf{X}(h) = \{X(nh), n \in Z^+\}$ of X, $h > 0$. Since $p_{jj}(nh) > 0$ for every $n \in Z^+$, every state of the Markov chain $\mathbf{X}(h)$ is aperiodic and therefore

$$\lim_{n\to\infty} p_{jk}(nh) = \alpha_{jk}(h) \quad , \quad \text{say,} \tag{3}$$

exists for all $j, k \in S$, $h > 0$. If h_1 and h_2 are two positive rational numbers, it is easy to verify that the sequences $\{nh_1, n \in Z^+\}$ and $\{nh_2, n \in Z^+\}$ have infinitely many points in common and these common points can be arranged as the increasing sequence $\{\nu_n, n \in Z^+\}$ converging to $+\infty$. Then by virtue of (3),

$$\alpha_{jk}(h_1) = \lim_{n\to\infty} p_{jk}(nh_1) = \lim_{n\to\infty} p_{jk}(\nu_n)$$

$$= \lim_{n\to\infty} p_{jk}(nh_2) = \alpha_{jk}(h_2) ;$$

i.e. $\alpha_{jk}(h)$ is independent of h for all rational $h > 0$. The result (1) follows by an appeal to the denseness of the rationals and the uniform continuity of $p_{jk}(t)$ on $\mathbb{R}^+$ guranteed by lemma 2.2.1. The result (2) is an immediate consequence of the Chapman-Kolmogorov equations. The proof is complete.

Suppose a state $j \in S$ of the discrete Markov process is persistent (transient) in the Markov chain $X(h)$ for some $h > 0$. Does it follow that it is persistent (transient) in $\mathbf{X}(h)$ for all $h > 0$? We can answer this question in the affirmative by using the following

Theorem 2 : The integral

$$\int_0^\infty p_{jj}(t)dt = \infty$$

iff $\Sigma p_{jj}(nh) = \infty$ for some $h > 0$, and in this case for all $h > 0$, $\Sigma p_{jj}(nh) = \infty$.

Proof : Let $h > 0$ and define

$$m_n(h) = \inf\{ p_{jj}(t) \mid nh \leq t \leq (n+1)h \},$$

$$M_n(h) = \sup\{ p_{jj}(t) \mid nh \leq t \leq (n+1)h \}, \quad n \in Z^+.$$

Observe that $m_0(h) = \inf\{ p_{jj}(t) \mid 0 \leq t \leq h \}$ is positive for every $h \in (0, \infty)$ in view of the continuity of $p_{jj}(t)$ on $[0, \infty)$ and the fact that $p_{jj}(t) > 0$ for every $t \in [0, \infty)$. By the Chapman-Kolmogorov equations

$$p_{jj}(t+s) \geq p_{jj}(t)\ p_{jj}(s)$$

so that for every $t \in [nh, (n+1)h]$,

$$p_{jj}(t) \geq p_{jj}(nh)\ p_{jj}(t - nh) \tag{4}$$

and

$$p_{jj}((n+1)h) \geq p_{jj}(t)\ p_{jj}((n+1)h - t)) . \tag{5}$$

One easy consequence of (4) and (5) is that

$$m_n(h) \geq m_0(h)\ p_{jj}(nh)$$

and

$$M_n(h) \leq [m_0(h)]^{-1}\ p_{jj}[(n+1)h] .$$

Hence for every $h > 0$ and N a positive integer

$$h\, m_0(h) \sum_{n=0}^{N-1} p_{jj}(nh) \leq h \sum_{n=0}^{N-1} m_n(h)$$

$$\leq \int_0^{Nh} p_{jj}(t)\, dt$$

$$\leq h \sum_{n=0}^{N-1} \hat{M}_n(h) \leq \{m_0(h)\}^{-1} \sum_{n=0}^{N-1} p_{jj}(nh) .$$

The statement of the theorem is an immediate consequence obtained by allowing $N \to \infty$ in the above string of inequalities.

In view of the above theorems 1 and 2, there is no ambiguity in the following definition.

Definition 1 : A state j of a discrete Markov process $\mathbf{X}$ is persistent (transient) iff it is persistent (transient) in a discrete skeletion $\mathbf{X}(h)$ of $\mathbf{X}$ for some $h > 0$. A persistent state j of $\mathbf{X}$ is a non-null (null) state iff it is non-null in some discrete skeleton $\mathbf{X}(h)$.

One can also easily prove the following

Theorem 3 : A state j of the discrete Markov process $\mathbf{X}$ is persistent (transient) according as the integral $\int_0^\infty p_{jj}(t)dt$ is $+\infty$ or finite. A persistent state j is non-null or null according as $\alpha_{jj} = \lim_{t\to\infty} p_{jj}(t)$ is positive or zero.

Corollary 1 : If $j \to k$ and k is a persistent state, then

$$\int_0^\infty p_{jk}(t)dt = +\infty.$$

Proof . This is an immediate consequence of the inequality

$$p_{jk}(s+t) \geq p_{jk}(s)\, p_{kk}(t)$$

obtained from the Chapman-Kolmogorov equations and the fact that $p_{jk}(t) > 0$ for all $t \in (0, \infty)$.

Corollary 2 . If k is transient, then for all $j \in S$,

$$\int_0^\infty p_{jk}(t)dt < \infty$$

and $\lim_{t\to\infty} p_{jk}(t) = 0$.

Proof : If $j \not\to k$, the result is trivial and if $j \leftrightarrow k$, the result follows by using the obvious inequality

$$p_{kk}(s+t) \geq p_{kj}(s)\, p_{jk}(t) .$$

Suppose now that $j \to k$ but $k \not\to j$. Then for every $h > 0$, $p_{jk}(nh) > 0$ and $p_{kj}(nh) = 0$, $n = 1, 2, \ldots$. Consider

$$\Pr[X(nh) \neq j,\ n = 1, 2, \ldots \mid X(0) = j]$$

$$\geq \Pr[X(h) = k,\ X(nh) \neq j,\ n = 2,3,\ldots \mid X(0) = j\,]$$

$$= p_{jk}(h) \Pr[X(nh) \neq j,\ n = 2, 3, \ldots \mid X(h) = k\,]$$

$$= p_{jk}(h) > 0 .$$

Thus $f_{jj}(h) < 1$ and therefore j is transient in $\mathbf{X}(h)$ for every $h > 0$. Thus j is transient in **X** also and the result of the corollary is an immediate consequence of the inequality

$$p_{jj}(s+t) \geq p_{jk}(t)\, p_{kj}(s)$$

or equivalently

$$p_{jk}(t) \leq p_{jj}(s+t)/p_{kk}(s) \ . \tag{6}$$

<u>Corollary 3</u> : If j is inessential, it is transient and a persistent state is essential.

<u>Theorem 4</u> : Being persistent is a solidarity property.

<u>Proof</u> : Let j be a persistent state and $C(j)$ be the class of states which communicate with j. If $C(j) = \{j\}$, there is nothing to be proved. So let $C(j) \neq \{j\}$ and $k \in C(j)$, $k \neq j$. That k is persistent follows from the inequality

$$p_{kk}(s+t+u) \geq p_{kj}(s)\, p_{jj}(t)\, p_{jk}(u)$$

by an application of theorem 3.

<u>Corollary 4</u> : Being transient, persistent non-null or persistent null are all solidarity properties.

<u>Corollary 5</u> : If k is persistent null, then $\lim_{t\to\infty} p_{jk}(t) = 0$ for all states $j \in S$.

<u>Proof</u> : If $j \nrightarrow k$, there is nothing to be proved. If $j \rightarrow k$ or $j \leftrightarrow k$, the result is a consequence of the inequality (6).

Theorem 5 : In a finite Markov process there are no null states and all its states can not be transient.

Proof : Suppose that the state-space of the finite Markov process is $\{1, 2, ..., M\}$. If possible let state 1 be persistent null and $C(1)$ be the corresponding class of states which communicate with state 1. Without loss of generality, we can take $C(1) = \{1, ..., m\}$, $m \leq M$. Since state 1 is persistent, by corollary 3, it is essential and therefore by theorem 3.3, $C(1)$ is minimal closed. In particular, for all $j \in C(1)$ and $t \geq 0$

$$\sum_{k=1}^{m} p_{jk}(t) = 1 .$$

We arrive at a contradiction by allowing $t \to \infty$ in view of the fact that $\lim_{t\to\infty} p_{jk}(t) = 0$ for all $j, k = 1, ..., m$, by virtue of corollary 5. If all states of the finite Markov process are transient, we arrive at a contradiction in the same manner.

We have discussed the classification of states of the discrete Markov process in terms of $p_{jk}(t)$. In many cases it is difficult to have explicit representation for $p_{jk}(t)$, which is amenable to easy algebraic treatment. It is therefore necessary to be able to obtain criteria for classifying states in terms of the intensity rates or of the $X(R)$ chain. The following theorem enables us to do this.

Theorem 6 : If a state j is persistent in any one of the processes $\mathbf{X}$, $\mathbf{X}(h)$, $\mathbf{X}(R)$ or $\mathbf{X}(P)$, it is persistent in all the other three processes.

Proof : It is easy to see from equations (2.3.6), (2.5) (2.6) that

$$\frac{1}{\beta} \sum_{n=0}^{\infty} p_{jj}^{(n)} = \int_0^{\infty} p_{jj}(t)dt = \sum_{m=0}^{\infty} \int_0^{\infty} p_{jj}(t,m)dt = \frac{1}{q_j} \sum_{n=0}^{\infty} r_{jj}^{(m)}$$

in the sense that either all are finite or all are infinite. The rest of the argument is trivial.

Recall that a state j of a Markov chain is persistent or transient according as the probability of at least one return to j is one or less than one. We now adopt this approach of basing the classification of states into persistent and transient states on the distribution of the first return times.

Let $\{ X(t), t \in IR^+ \}$ be a seperable discrete Markov process on a complete probability space (Ω, IF, P). Let $s \in IR^+$ and $X(s) = j \in S$. Define

$$\rho_j = \inf\{ t \mid t > 0, X(s+t) \neq j, X(s) = j \}$$

as the exit time from state j and let

$$\tau_{jk} = \inf \{t \mid t > \rho_j , X(t) = k \}$$

denote the first entrance time in state k from state j. It is easy on account of the separability of the process to see that ρ_j and τ_{jk} are both r.v.s. Let

$$F_{jk}(t) = Pr[\tau_{jk} \leq t] , \quad 0 \leq t < \infty, \tag{7}$$

which is in general a defective distribution function in the sense that $F_{jk}(+\infty) = \lim_{t\to\infty} F_{jk}(t)$ may be less than one. We are interested in $F_{jj}(.)$, which we define by (7) with k replaced by j and in the particular case when j is an absorbing state, we take

$$F_{jj}(t) = \begin{cases} 0, & t < 0, \\ 1, & t \geq 0 \end{cases}$$

It is obvious that $F_{jj}(+\infty)$ is the probability of at least one return of the process to its initial state j. In analogy with Markov chains we can introduce the following

<u>Definition 2</u> : A state j of a discrete Markov process is persistent (transient) if $F_{jj}(+\infty)$ is one (less than one).

It is now necessary to demonstrate that this definition does not contradict the earlier definition 1. We need the following two theorems on Laplace transforms and Laplace - Stieltjes transforms for the proof of which we refer to Widder (1946).

<u>Abelian theorem (A) for Laplace transforms</u> :

(i) If $\int_0^\infty a(t)dt$ is convergent, then $\int_0^\infty e^{-ut}a(t)dt$ is uniformly convergent for the real part Re(u) of u positive. Moreover,

$$\lim_{u\to 0} \int_0^\infty e^{-ut}\, a(t)dt = \int_0^\infty a(t)dt ,$$

provided $u \to 0$ such that $|\arg u| \leq \theta < \pi/2$.

(ii) if $a(t) \geq 0$, $t \in \mathbb{R}^+$, and $\int_0^\infty e^{-ut} a(t)dt < \infty$ for $Re(u) > 0$, then

$$\lim_{u \to 0} \int_0^\infty e^{-ut} a(t)dt = \int_0^\infty a(t)\, dt, \quad |\arg (u)| \leq \theta < \pi/2 ,$$

in the sense that both the sides are either finite and equal or both are $+\infty$.

(iii) If $\int e^{-ut} a(t)dt < \infty$ for $Re(u) > 0$ and $a(t)$ has a limit as $t \to \infty$, then

$$\lim_{t \to \infty} a(t) = \lim_{u \to 0} u \int_0^\infty e^{-ut} a(t)\, dt, \quad |\arg(u)| \leq \theta < \pi/2.$$

Abelian theorem (B) for Laplace-Stieltjes Transform :

If $a(t)$, $t \in \mathbb{R}^+$ is of bounded variation in every finite interval, and if $\int_0^\infty e^{-ut} da(t) < \infty$ for $Re(u) > 0$, and if $\lim_{t \to \infty} a(t)$ exists, then

$$\lim_{t \to \infty} a(t) = \lim_{u \to 0} \int_0^\infty e^{-ut} da(t) .$$

Lemma 1 : For a discrete Markov process

$$p_{jk}(t) = \exp[-q_j t]\, \delta_{jk} + \int_0^t p_{jk}(t-u)\, dF_{jk}(u) . \tag{8}$$

Proof : Let $j \neq k$ and observe that

$$p_{jk}(t) = \Pr[X(t) = k \mid X(0) = j]$$

$$= E[E\{I[X(t) = k] \mid \tau_{jk}, X(0)\} \mid X(0) = j]$$

$$= \int_0^t E[I[X(t)=k] \mid \tau_{jk} = u, X(0) = j]\, dF_{jk}(u)$$

$$= \int_0^t E[I[X(t) = k] \mid X(u) = k]\, dF_{jk}(u)$$

$$= \int_0^t p_{kk}(t-u)\, dF_{jk}(u) .$$

When $j = k$, a similar argument yields (8), provided we note that the process may continue in state j during $[0, t]$ and that this happens with probability $\exp(-q_j t)$ by virtue of theorem 2.6.2.

Lemma 2 : If

$$G_{jk}(u) = \int_0^\infty e^{-ut} p_{jk}(t)dt \quad \text{and} \quad H_{jk}(u) = \int_0^\infty e^{-ut} dF_{jk}(t) ,$$

then

$$G_{jk}(u) = (u+q_j)^{-1} \delta_{jk} + G_{kk}(u) H_{jk}(u), \quad u > 0 . \tag{9}$$

Proof : This is a straight forward consequence of relation (8).

Theorem 7 : Definitions 1 and 2 of a persistent (transient) state are equivalent for a finite Markov process.

Proof : Suppose j is transient according to definition 2 i.e., suppose $F_{jj}(+\infty) < 1$. Then we have from (9), that

$$G_{jj}(u) = (u + q_j)^{-1} \{1 - H_{jj}(u)\}^{-1}$$

and therefore by Abelian theorem (B), $\lim_{u \to 0} H_{jj}(u) < 1$ and thus $\lim_{u \to 0} G_{jj}(u) < \infty$. But by Abelian theorem (A),

$$\lim_{u \to 0} G_{jj}(u) = \int_0^\infty p_{jj}(t)\, dt.$$

Hence by theorem 3, j is transient according to definition 1.

Conversely, if j is transient according to definition 1, then $\int_0^\infty p_{jj}(t)\, dt < \infty$, so that by Abelian theorem (A), $\lim_{u \to 0} H_{jj}(u) < 1$, Thus by Abelian theorem (B),

$$\lim_{t \to \infty} F_{jj}(t) = \lim_{u \to 0} H_{jj}(u) < 1.$$

which completes the proof.

5. ASYMPTOTIC PROPERTIES

We have already seen that according to theorem 4.5 a finite Markov process has at least one ergodic state and may have some transient states. One may therefore assume, without much loss of generality, that in our finite Markov process with state-space $S = \{1, \ldots, M\}$, the first $m < M$ states $1, 2, \ldots, m$ form a minimal closed class C of ergodic state and that the remaining $M-m$ states form a class $T = \{m+1, \ldots, M\}$ of inessential and therefore transient states. Every state $j \in T$ leads to every state $k \in C$ but no state in C leads to any state in T. In this section we shall not consider any other type of finite Markov process. According to theorem 4.1 $\alpha_{jk} = \lim_{t\to\infty} p_{jk}(t)$ exists for all $j, k \in S$. Under the structure described above, we shall now demonstrate that $p_{jk}(t)$ attains its limit at an exponential rate.

Theorem 1 : Let $\{X(t), t \in \mathbb{R}^+\}$ be a finite Markov process with state-space $S = \{1, \ldots, m, m+1, \ldots, M\}$ such that $C = \{1, \ldots, m\}$ is an ergodic class and $T = \{m+1, \ldots, M\}$ is a transient class, every state of which leads to all states in C. Then $\alpha_{jk} = \lim_{t\to\infty} p_{jk}(t) = \alpha_k$ is independent of j for all $k \in S$. Moreover, there exist non-negative numbers α and ρ such that

$$\text{(i)} \quad | p_{jk}(t) - \alpha_k | \leq \alpha \rho^t , \; t > 0 ;$$

$$\text{(ii)} \quad \alpha_k > 0, \; k \in C, \quad \alpha_k = 0 , \; k \in T \text{ and } \sum_{k\in C} \alpha_k = 1.$$

Proof : Define for $t > 0$ and $k \in S$,

$$\lambda_k(t) = \max \{p_{jk}(t) \mid j \in S \}, \quad \mu_k(t) = \min \{p_{jk}(t) \mid j \in S\}. \quad (1)$$

One can easily verify by using the Chapman-Kolmogorov equations that for each fixed $k \in S$, $\lambda_k(.)$ is non-increasing and $\mu_k(.)$ is non-decreasing on $(0, \infty)$ so that, they being bounded functions, converge to finite limits as $t \to \infty$. We claim that

$$\lim_{t\to\infty} \lambda_k(t) = \lim_{t\to\infty} \mu_k(t) = \alpha_k , \quad k \in S . \tag{2}$$

Since all states $j \in S$ lead to every state $k \in C$, $p_{jk}(t)>0$ for all $t > 0$. Let $h > 0$ be a fixed number and define

$$\delta(h) = \min \{p_{jk}(h) \mid j \in S , \quad k \in C \}$$

which is obviously positive. Let a and b be two fixed states in S and define

$$S_1 = \{ k \mid k \in S, \ p_{ak}(h) \geq p_{bk}(h)\} , \qquad S_2 = S - S_1 .$$

Then

$$\sum_{k\in S_1} \{p_{ak}(h) - p_{bk}(h)\} + \sum_{k\in S_2} \{p_{ak}(h) - p_{bk}(h)\} = 0 .$$

Hence, if s denotes the number of ergodic states in S_1,

$$\sum_{k\in S_1} p_{ak}(h) - \sum_{k\in S_1 \cap T} p_{bk}(h)$$

$$= \sum_{k\in S_1 \cap C} p_{bk}(h) + \sum_{k\in S_2} p_{bk}(h) - \sum_{k\in S_2} p_{ak}(h)$$

$$\leq 1 - \sum_{k\in S_2 \cap C} p_{ak}(h) \leq 1 - (m-s)\delta . \tag{3}$$

Moreover, $$\sum_{k\in S_1 \cap C} p_{bk}(h) \geq s\delta . \tag{4}$$

Combining the inequalities (3) and (4), one has

$$\sum_{k\in S_1} \{p_{ak}(h) - p_{bk}(h)\} \leq (1 - m\delta)$$

which is an upper bound not depending on a and b. Thus

$$\max\{\sum_{k\in S_1} \{p_{ak}(h) - p_{bk}(h)\} \mid a, b \in S\} \leq 1 - m\delta \,. \tag{5}$$

Observe now that

$$\begin{aligned} \lambda_k(h) - \mu_k(h) &= \max\{p_{ak}(h) - p_{bk}(h) \mid a, b \in S\} \\ &\leq \max\{\sum_{k\in S_1} \{p_{ak}(h) - p_{bk}(h)\} \mid a, b \in S\} \\ &\leq (1 - m\delta) \,. \end{aligned} \tag{6}$$

Let $t > 0$ and consider

$$\lambda_k(t+h) - \mu_k(t+h)$$

$$= \max_{a,b\in S} [\sum_{j\in S_1}\{p_{aj}(h)-p_{bj}(h)\}p_{jk}(t) + \sum_{j\in S_2}\{p_{aj}(h)-p_{bj}(h)\}p_{jk}(t)]$$

$$\leq \max_{a,b\in S} [\sum_{j\in S_1}\{p_{aj}(h) - p_{bj}(h)\}\{\lambda_k(t) - \mu_k(t)\}]$$

$$\leq \{\lambda_k(t) - \mu_k(t)\} (1 - m\delta) \,. \tag{7}$$

Using (6) and (7) repeatedly it follows that

$$\lambda_k(nh) - \mu_k(nh) \leq (1 - m\delta)^n, \qquad n = 1, 2, \ldots . \tag{8}$$

Observe that by definition, $m\delta \leq 1$. If $m\delta = 1$, one can easily verify that $\lambda_k(nh) = \mu_k(nh)$ for all $n \geq 1$ and therefore $\lambda_k(t) \equiv \mu_k(t)$. If $m\delta < 1$, $(1 - m\delta)^n \to 0$ as $n \to \infty$ and therefore by virtue of (8)

$$\lim_{t\to\infty} \lambda_k(t) = \lim_{t\to\infty} \lambda_k(nh) = \lim_{t\to\infty} \mu_k(nh) = \lim_{t\to\infty} \mu_k(t) ,$$

which completes the proof of the first part of our claim (2). Let α_k be the common limit of the functions $\lambda_k(\cdot)$ and $\mu_k(\cdot)$. Then by definition, for every $t > 0$,

$$\lambda_k(t) \geq \alpha_k \geq \mu_k(t)$$

and therefore,

$$| \, p_{jk}(t) - \alpha_k \, | \leq \lambda_k(t) - \mu_k(t) \to 0 \quad \text{as} \quad t \to \infty .$$

i.e., $\lim p_{jk}(t) = \alpha_k$, and the claim (2) is proved.

To obtain (i), write $t = \nu h + r$, $\nu \geq 0$, $0 \leq r < h$, and let $[\nu]$ denote the integral part of ν. Observe that by virtue of (7)

$$|p_{jk}(t) - \alpha_k| \leq \lambda_k(t) - \mu_k(t)$$

$$\leq (1 - m\delta)^{\nu} \{\lambda_k(r) - \mu_k(r)\}$$

$$\leq (1 - m\delta)^{[t/h]-1}$$

$$\leq \alpha\rho^t$$

with $\alpha = (1 - m\delta)^{-1}$, $\rho = (1 - m\delta)^{1/h}$.

The first part of assertion (ii) is an easy consequence of the fact that $\alpha_k \geq \mu_k(t)$ for every $t > 0$ and therefore for $k \in C$

$$\alpha_k \geq \min \{p_{jk}(h) \mid j \in S\}$$

$$\geq \min \{p_{jk}(h) \mid j \in S , k \in C\} = \delta > 0.$$

If $k \in T$, $p_{jk}(t) \to 0$ as $t \to \infty$ by virtue of corollary 4.2. The last part of (ii) follows immediately. The theorem is proved.

<u>Corollary 1</u> : If $X(0) = j \in T$ with positive probability, then the Markov process $\{X(t), t \in IR^+\}$ is eventually absorbed in C with probability one. The probability that it remains for ever in T is zero.

<u>Proof</u> : Let $E(t) = [X(t) \in C]$ denote the event that the process is in some state in C at epoch t and let $E = \bigcup_{t>0} [X(t) \in C]$ denote the set that $X(t) \in C$ for some epoch $t > 0$. Since $E(t) \subset E(t+h)$ for any $h > 0$, it is readily verified that for any sequence $\{t_n\}$, $0 < t_n \uparrow \infty$, as $n \to \infty$,

$$E = \lim_{n\to\infty} E(t_n) = \bigcup_{n=1}^{\infty} E(t_n)$$

and therefore E is an event. It also follows that

$$\Pr[E \mid X(0) = j] = \lim_{n\to\infty} \Pr[E(t_n) \mid X(0) = j]$$

$$= \lim_{n\to\infty} \sum_{k\varepsilon C} p_{jk}(t_n)$$

$$= \sum_{k\varepsilon C} \alpha_k = 1.$$

The rest of the argument is trivial.

<u>Corollary 2</u> : The Markov process $\{X(t),\ t \ \varepsilon\ \mathrm{IR}^+\}$ satisfying the conditions of the above theorem 1, converges in distribution to a r.v. Z such that

$$\Pr[Z = k] = \alpha_k, \quad k \ \varepsilon\ S,$$

for every initial distribution.

<u>Proof</u> : It is an immediate consequence of the fact that for any $t > 0$

$$\Pr[X(t) = k] = \sum_{j\varepsilon S} \Pr[X(0) = j]\, p_{jk}(t)$$

$$\to \alpha_k, \quad k \ \varepsilon\ S$$

by virtue of theorem 1.

In view of the result of this corollary, the probability distribution specified by $(\alpha_1, \ldots, \alpha_m, 0, \ldots, 0)$ is sometimes referred to as the stationary distribution or the limiting distribution. We shall investigate the properties of this distribution in the next section.

6. INVARIANT DISTRIBUTION

Suppose that a finite Markov process $\{X(t), t \in \mathbb{R}\}$ on the state-space $S = \{1, \ldots, M\}$ is irreducible. Then according to theorem 4.4, all its states are ergodic and further by theorem 5.1 $\alpha_k = \lim_{t\to\infty} p_{jk}(t)$ is positive and independent of j for all $k \in S$. The probability distribution specified by $\{\alpha_1, \ldots, \alpha_M\}$ is also the limit distribution of the process as $t \to \infty$. This probability distribution has another interesting interpretation in the light of the following definition of a strictly stationary stochastic process.

Definition 1 : A stochastic process $\{X(t), t \in T\}$, where $T = \mathbb{R}^+$ or Z^+ is a strictly stationary process if for all $n \geq 1$ $t_1, \ldots, t_n \in T$, $h > 0$ such that $t_j+h \in T$, $j = 1, \ldots, n$, the joint distribution of $X(t_1), \ldots, X(t_n)$ is the same as that of $X(t_1+h), \ldots, X(t_n+h)$.

If the state-space of the process is countable, the above definition is equivalent to the requirement that

$$\Pr[X(t_1) = j_1, \ldots, X(t_n) = j_n] = \Pr[X(t_1+h) = j_1, \ldots, X(t_n+h) = j_n] \quad (1)$$

for all $n \geq 1$, $t_1, \ldots, t_n \in T$, $h > 0$ such that $t_r+h \in T, r=1, \ldots, n$.

This condition for strict stationarity further simplifies to a single condition on the initial distribution in the case of a discrete Markov process as established in the following

Lemma 1 : A discrete Markov process is a strictly stationary process iff the initial distribution $\pi_j = \Pr[X(0) = j]$, $j \varepsilon S$ is such that for all $t > 0$

$$\pi_k = \sum_{j\varepsilon S} \pi_j \, p_{jk}(t), \qquad k \varepsilon S . \tag{2}$$

Proof : Observe that if $0 < t_1 <\ldots< t_n < \infty$, then

$$\Pr[X(t_1) = k_1] = \sum_{j\varepsilon S} \Pr[X(0) = j] \, p_{jk_1}(t_1) \tag{3}$$

and

$$\Pr[X(t_1) = k_1,\ldots,X(t_n)=k_n] = \Pr[X(t_1)=k_1] \prod_{r=2}^{n} p_{k_{r-1}k_r}(t_r-t_{r-1}). \tag{4}$$

It follows from (3) that if the discrete Markov process is strictly stationary then (2) must hold. Conversely if (2) holds, then by virtue of (3) and (4), the process is a strictly stationary process.

This lemma suggests the following

Definition 2 : A probability distribution $\{\pi_j, j \varepsilon S\}$ is an invariant distribution or a stationary distribution associated with a discrete Markov process $\{X(t), t \varepsilon \mathbb{R}^+\}$ on S iff (2) holds for all $t > 0$.

Example 1 : Let $\{X(t), t \varepsilon \mathbb{R}^+\}$ be the two-state Markov process defined in example 1.3.3 with

$$p_{00}(t) = \mu/(\lambda+\mu) + [\lambda \exp\{-(\lambda+\mu)t\}]/(\lambda+u), \quad p_{01}(t) = 1 - p_{00}(t)$$

$$p_{11}(t) = \lambda/(\lambda+\mu) + [\mu \exp\{-(\lambda+\mu)t\}]/(\lambda+\mu), \quad p_{10}(t) = 1 - p_{11}(t)$$

It is easy to verify that equations (2) have the unique solution

$$\pi_0 = \mu/(\lambda+\mu) = \alpha_0 \quad , \quad \pi_1 = \lambda/(\lambda+\mu) = 1 - \pi_0 = \alpha_1$$

which is then the unique invariant distribution. Note that $\pi_0 = \alpha_0$, $\pi_1 = \alpha_1$; which is not an accidental result as established in the following theorem.

Theorem 1 : A finite, irreducible Markov process has the unique invariant distribution

$$\pi_k = \alpha_k = \lim_{t\to\infty} p_{jk}(t) \quad , \quad k \in S . \tag{5}$$

Proof : The finiteness and irreducibility of the process imply that its states are all ergodic and therefore

$$\alpha_k = \alpha_{jk} = \lim_{t\to\infty} p_{jk}(t) \ , \quad j \in S, \quad k \in S$$

by virtue of theorem 5.1. Thus by theorem 4.1, $\{\alpha_k, k \in S\}$ is a solution of (2), and therefore an invariant distribution of the process.

If possible, let $\{\pi_k, k \in S\}$ be any other solution of (2) . Then

$$\pi_k = \sum_{k \varepsilon S} \pi_j \, p_{jk}(t) \rightarrow \alpha_k \qquad \text{as} \quad t \rightarrow \infty$$

and therefore $\pi_k = \alpha_k$, $k \varepsilon S$, which concludesthe proof of the theorem.

Suppose that a finite Markov process is reducible and that it has two minimal closed classes $C_1 = \{1, \ldots, m_1\}$, and $C_2 = \{m_1+1, \ldots, m_1 + m_2\}$, say, whose union is S. It is easy to verify by employing the above theorem that $\{\alpha_1, \ldots, \alpha_{m_1}, 0, \ldots, 0\}$ and $\{0, \ldots, 0, \alpha_{m_1+1}, \ldots, \alpha_{m_1+m_2}\}$ are both invariant distribu - tions associated with the reducible Markov process. In fact if a discrete Markov process has two (or more) distinct invariant distributions, $\{\pi_j, j \varepsilon S\}$ and $\{\nu_j, j \varepsilon S\}$, then it has infinitely many invariant distributions because for $\alpha \varepsilon [0, 1]$, $\{\alpha\pi_j + (1-\alpha)\nu_j, j \varepsilon S\}$ is also an invariant distribution.

In general, it would be difficult to use equations (2) to obtain the invariant distribution in view of the algebraic diffi - culties in obtaining $p_{jk}(t)$ from a knowledge of the intensity rates q_{jk} only. The following theorem resolves this difficulty.

Theorem 2 : The invariant distribution $\{\pi_j, j \varepsilon S\}$ is a solution of the equations

$$\sum_{j \varepsilon S} \pi_j \, q_{jk} = 0, \quad k \varepsilon S. \tag{6}$$

Proof : By definition

$$\pi_k = \Sigma \pi_j p_{jk}(t)$$

and therefore for all $t > 0$

$$\pi_k \{p_{kk}(t) - 1\} + \sum_{j \in S-\{k\}} \pi_j p_{jk}(t) = 0 . \tag{7}$$

One obtains (6) by allowing $t \downarrow 0$ in (7) after dividing its both sides by t.

Before giving examples of invariant distribution, we discuss the connection between invariant distributions of the $\mathbf{X} = \{X(t), t \in \mathbb{R}^+\}$ - process and those of the Markov chains $\mathbf{X}(h)$, $\mathbf{X}(P)$ and $\mathbf{X}(R)$ respectively.

Theorem 3 : Let $\mathbf{X} = \{X(t), t \in \mathbb{R}^+\}$ be an irreducible finite Markov process. Then the Markov chains $\mathbf{X}(h)$ and $\mathbf{X}(P)$ have the same invariant distribution as that of $\mathbf{X}$. Moreover, if $\{\beta_k, k \in S\}$ is the invariant distribution of $\mathbf{X}(R)$, then

$$\beta_k = \alpha_k q_k / \{ \sum_{j \in S} \alpha_j q_j \} , \quad k \in S. \tag{8}$$

Proof : Let $\{\alpha_k, k \in S\}$ be the unique invariant distribution of the finite Markov process $\mathbf{X}$. Then

$$\alpha_k = \Sigma \alpha_j p_{jk}(t) , \quad k \in S ,$$

for all $t > 0$ and in particular for $t = h$. Hence it is also the invariant distribution of $\mathbf{X}(h)$.

Moreover, $\sum_{j\in S} \alpha_j q_{jk} = 0$, $k \in S$ and $P = I + \beta^{-1} Q$, imply that

$$\Sigma \alpha_j p_{jk} = \sum_{j\in S} \alpha_j \{\delta_{jk} + \beta^{-1} q_{jk}\} = \alpha_k$$

and therefore $\{\alpha_k, k \in S\}$ is also the invariant distribution of $\mathbf{X}(P)$.

Finally let $\{\beta_k, k \in S\}$ be the invariant distribution of $\mathbf{X}(R)$ with transition matrix

$$R = I + D^{-1} Q ,$$

where $D = \text{diag}\{q_1, \ldots, q_M\}$. Then for all $k \in S$

$$\beta_k = \sum_{j\in S} \beta_j r_{jk} = \sum_{j\in S-\{k\}} \beta_j q_{jk}/q_j$$

$$= \sum_{j\in S} \{\beta_j q_{jk}/q_j\} + \beta_k$$

so that $\sum_{j\in S} \beta_j q_{jk}/q_j = 0$, and thus

$$\beta_k/q_k = c\,\alpha_k , \quad c \neq 0 , \quad k \in S ,$$

by virtue of theorems 1 and 2. The requirement that $\Sigma \alpha_k = 1$ implies that $c = 1/\{\sum_{j\in S} \alpha_j q_j\}$, which yields (8) and the proof is complete.

7. INVARIANT DISTRIBUTION : BIRTH-DEATH PROCESSES

The most commonly used finite Markov process in modelling a variety of real life phenomena is the birth-death process on the set $S = \{0, 1, \ldots, M\}$ or a subset of S with parameters λ_j and μ_j, $j \in S$ as given by the definition 2.5.2. The following lemma provides a general solution which is specialized to particular cases in the examples that follow :

Lemma 1 : Let $\{X(t), t \in \mathbb{R}^+\}$ be a finite birth-death process on the set $\{0, 1, \ldots, M\}$ with positive birth rates $\lambda_0, \lambda_1, \ldots, \lambda_{M-1}$ and positive death rates $\mu_1, \ldots, \mu_M$. Its invariant distribution $\{\pi_k, k \in S\}$ is specified by

$$\pi_k = \prod_{j=1}^{k} \{\lambda_{j-1}/\mu_j\} \pi_0, \quad k = 1, \ldots, M, \tag{1}$$

where

$$\pi_0 = \{1 + \sum_{k=1}^{M} \prod_{j=1}^{k} (\lambda_{j-1}/\mu_j)\}^{-1} . \tag{2}$$

Proof : Since $\lambda_0, \ldots, \lambda_{M-1}$ and $\mu_1, \ldots, \mu_M$ are all positive, the birth-death process is an irreducible ergodic Markov process. Its invariant distribution $\{\pi_k, k \in S\}$ is specified by the equations (6.6) which are

$$-\lambda_0 \pi_0 + \mu_1 \pi_1 = 0 ,$$

$$\lambda_{k-1} \pi_{k-1} - (\lambda_k + \mu_k)\pi_k + \mu_{k+1} \pi_{k+1} = 0, \quad k = 1, \ldots, M-1, \tag{3}$$

$$\lambda_{M-1} \pi_{M-1} - \mu_M \pi_M = 0 .$$

Define $\xi_k = -\lambda_k \pi_k + \mu_{k+1} \pi_{k+1}$, $k = 0, 1, \ldots, M-1$ and observe that $\xi_0 = 0$, $\xi_{M-1} = 0$ and

$$\xi_k - \xi_{k-1} = 0 , \quad k = 1 , \ldots, M-2 .$$

Hence $\xi_k = 0$ for all $k = 0 , 1, \ldots, M-1$. In particular

$$\pi_k = (\lambda_{k-1}/\mu_k)\pi_{k-1} , \quad k = 1, \ldots, M$$

so that

$$\pi_k = \{ \prod_{j=1}^{k} (\lambda_{j-1}/\mu_j)\}\pi_0, \quad k = 1, \ldots, M .$$

The expression (2) for π_0 is a consequence of the fact that

$$\sum_{j=0}^{M} \pi_j = 1.$$

<u>Example 1</u> : Random walk with reflecting barriers [cf. Example 2.5.1] . Here $\lambda_0 = \ldots = \lambda_{M-1} = \lambda$ and $\mu_1 = \ldots = \mu_M = \mu$, so that

$$\pi_k = (\lambda/\mu)^k \pi_0 , \quad k = 1, \ldots, M ,$$

and

$$\pi_0 = \begin{cases} (1-\delta)/(1-\delta^{M+1}) , & \text{if } \delta = \lambda/\mu \neq 1 , \\ 1/(M+1) , & \text{if } \delta = 1. \end{cases}$$

Example 2 : Elephant herds [cf. Example 2.5.3] .

Here $S = \{1, \ldots, M\}$, $\lambda_j = (M-j)\lambda$, $j = 1, \ldots, M-1$ and $\mu_j = (j-1)\mu$, $j = 2, \ldots, M$. Thus, one has after some simplification,

$$\pi_k = \binom{M-1}{k-1} \left(\frac{\lambda}{\lambda+\mu}\right)^{k-1} \left(\frac{\mu}{\lambda+\mu}\right)^{m-k+1}, \quad k = 1, \ldots, M ,$$

which is in fact the probability mass function of the r.v. $X+1$ where X has binomial distribution $B(n, p)$, with $n = M-1$, $p = \lambda/(\lambda+\mu)$.

Example 3 : Electric welders [cf. Example 2.5.4] .

Here $S = \{0, 1, \ldots, M\}$ and $\lambda_j = (M-j)\lambda$, $\mu_j = j\mu$, $j = 0, \ldots, M$. A comparison with example 2 above or equation 2.5.37 immediately yield

$$\pi_j = \binom{M}{j} \{\lambda/(\lambda+\mu)\}^j \{\mu/(\lambda+\mu)\}^{M-j}, \quad j = 0,1,\ldots,M .$$

Example 4 : Machine-repairmen problem [cf. Feller(1972)p.462].

The Swedish investigator Palm introduced a Markov model to study the efficient utilization of a large number M of identical machines looked after by $r \leq M$ operators. Thus, for example, in the weaving industry, a single operator may be asked to look after the working of a more than one loom. Our assumptions are as follows.

Suppose that each repairman can service only one machine at

any given epoch and that only one repairman looks after a machine under repair. The machines break down independently and the repairman also work independently. The probability that a machine working at epoch t needs attention before t+h is $\lambda h + o(h)$. The probability that a machine in repair at epoch t will be functioning again before t+h is $\mu h + o(h)$. Here λ and μ are parameters which do not depend on t or on the number of machines being repaired. If a machine fails and a repairman is available, he immediately attends to the machine, otherwise the machine awaits repair until a repariman is free to attend to it.

Let $X(t)$ denote the number of machines not working at epoch t. It is easy to see that $\{X(t), t \in \mathbb{R}^+\}$ is a birth-death process on the state-space $S = \{0, 1, \ldots, M\}$ with

$$\lambda_k = (M-k)\lambda \quad , \quad k = 0, 1, \ldots, M-1 ,$$

$$\mu_k = \begin{cases} k\,\lambda & , \quad k = 1, \ldots, r \quad , \\ r\,\mu & , \quad k = (r+1), \ldots, M. \end{cases}$$

One can now obtain the following invariant distribution by using equations (1) and (2) ;

$$\pi_k = \binom{n}{k} (\lambda/\mu)^k \pi_0 \quad , \quad 1 < k < r \quad ,$$

$$\pi_k = \binom{n}{k} (\lambda/\mu)^k \pi_0/r^{k-r} \quad , \quad r < k \leq M \quad ,$$

where

$$\pi_0 = \{ \sum_{k=0}^{r} \binom{n}{k} (\lambda/\mu)^k + \sum_{k=r+1}^{M} \binom{n}{k} (\lambda/\mu)^k/r^{k-r} \}^{-1} .$$

Example 5 : Queues with losses [cf. Gnedenko and Kovalenko (1968), p. 25] .

Suppose that at a servicing facility, there are M servers. If a server is free at the epoch of arrival of a customer, he is served immediately. The service facility has a waiting room which can accommodate at most m waiting customers. An arriving customer waits if all servers are busy and the number of waiting customers is less than m. If m customers are already waiting the arriving customer is lost to the service facility.

Suppose that the probability that a customer arriving during $(t, t+h]$ is $\lambda h + o(h)$. The service going on at epoch t is completed before t+h with probability $\mu h + o(h)$. If $X(t)$ denotes the total number of custoerms in the system, i.e. those being served and those who wait, then once again we have the birth-death process $\{X(t), t \in IR^+\}$ on the state-space $S = \{0,1,\ldots, M+m\}$ and with parameter

$$\lambda_k = \lambda \quad , \qquad k = 0, 1, \ldots, M+m-1 ,$$

$$\mu_k = \begin{cases} k\mu & , k = 1,\ldots, M, \\ M\mu & , k = M+1, \ldots, M+m . \end{cases}$$

The invariant distribution is given by

$$\pi_k = \begin{cases} \rho^k \pi_0/k! \quad , & 1 \le k \le M, \\ \rho^k \pi_0/\{M!\, M^{k-M}\}, & M \le k \le M+m , \end{cases}$$

where $\rho = \lambda/\mu$, and

$$\pi_0 = [\sum_{k=0}^{M} \beta^k/k! + \{\rho^m/M!\} \sum_{k=1}^{m} (\rho/M)^k]^{-1} .$$

CHAPTER IV

STATISTICAL PROPERTIES

1. INTRODUCTION

An important aspect of the study of a finite Markov process is the investigation of the statistical properties of the random variables or statistics associated with its evolution in the light of discussion in section 2.6.

Suppose $X(0) = i_0$ which is not an absorbing state. The process stays in state i_0 for a random duration T_{i_0} which has exponential distribution with scale parameter q_{i_0}. At the end of this sojourn in i_0, the process moves to state i_1 with probability $q_{i_0 i_1}/q_{i_0}$. If i_1 is also a non-absorbing state, the process stays in i_1 for an exponentially distributed random time T_{i_1} after which it moves to a new state i_2 with probability $q_{i_1 i_2}/q_{i_1}$ and so on. In order to formulate this description of evolution of a finite Markov process $\{X(t), t \in \mathbb{R}^+\}$ on a probability space $(\Omega, \mathbb{F}, P)$ more precisely, we introduce some terminology and notation.

A direct transition from state i to state j, $i \neq j$, is said to occur at epoch t if $X(t) = j$ and there exists an $\varepsilon > 0$ such that for all $u \in [t - \varepsilon, t)$, $X(u) = i$.

We then also say that the process has jumped from state i to state j at epoch t. Let $\tau_0 = 0, \tau_1, \tau_2, ..$ denote the random epochs of the jumps of a finite Markov process. They are formally defined in a recursive manner as follows.

Let Ω^* denote the subset of Ω, $P(\Omega^*) = 1$ such that for every $w \in \Omega^*$, the sample function $\{X(t, w), t \in \mathbb{R}^+\}$ is a step function. Define for $w \in \Omega^*$, $\tau_0(w) = 0$ and let

$$\tau_1(w) = \begin{cases} \inf\{u \mid u > 0, \ X(u, w) \neq X(0, w)\} \\ \infty, \quad \text{if } X(u, w) = X(0, w) \\ \qquad \text{for all } u \in \mathbb{R}^+ . \end{cases} \tag{1}$$

For $n > 1$,

$$\tau_{n+1}(w) = \begin{cases} \inf\{u \mid \tau_n(w) < u < \infty, X(u,w) \neq X(\tau_n(w), w)\} \\ \infty \quad \text{if } \tau_n(w) = +\infty \quad \text{or} \quad X(u,w) = X(\tau_n(w),w) \\ \qquad \text{for all } u > \tau_n(w) . \end{cases} \tag{2}$$

The functions τ_n can be defined arbitrarily for $w \in \Omega - \Omega^*$, without affecting their statistical properties. It is obvious that the sequence $\{\tau_n, n \in Z^+\}$ is an almost surely non-decreasing sequence. Define therefore,

$$N(t) = \max\{ n \mid n \in Z^+, \ \tau_n \leq t \} \tag{3}$$

which is the number of jumps of the process in $(0, t]$. Let

$$T_n = \begin{cases} \tau_{n+1} - \tau_n , & \text{if } \tau_n < \infty, \\ \infty , & \text{if } \tau_n = \infty ; \end{cases} \tag{4}$$

which is the duration between the n-th and (n+1)-th jump of the Markov process. If X_0 represents the initial state $X(0)$ and $X_n = X(\tau_n)$, the state reached after the n-th jump, $n \geq 1$, then T_n is the sojourn time in state X_n.

One can employ standard seperability arguments to demonstrate that the functions τ_n, $N(t)$, T_n and X_n, defined on $(\Omega, \mathbb{F}, P)$, are random variables, some of which may be extended real-valued. It follows that almost every observation of the process on $[0, t]$ is expressible as the random vector

$$V_t = \{X_0, T_0, X_1, T_1, \ldots, X_{N(t)-1}, T_{N(t)-1}, X_{N(t)}\} \tag{5}$$

of random dimension $2N(t) - 1$ if $N(t) > 0$ and as $\{X_0\}$ if $N(t) = 0$.

Suppose we have observed a finite Markov process on the interval $[0, t]$. Then

$$N_t(i,j) = \sum_{r=0}^{N(t)-1} I[X_r = i, X_{r+1} = j]\, I[N(t) > 0] \tag{6}$$

is the number of direct jumps from i to j during $[0, t]$ and is refered to as a <u>transition count</u>.

Let

$$A_t(i) = \int_0^t I[X(u) = i]du \tag{7}$$

be the total time spent by the process in state i during $[0,t]$. It is refered to as the sojourn time in state i during $[0, t]$. The integral in (7) is defined as a point-wise integral in the sense that $A_t(i)$ is a function on Ω with value

$$A_t(i)(w) = \int_0^t I[X(u) = i](w)\, du .$$

The statistics $N_t(i, j)$, $i \neq j$ and $A_t(i)$, $i, j \in S$ play a very important role in the problems of statistical inference for a finite Markov process. We obtain their first two moments in section 2 [cf. Albert (1962)]. In section 3 we describe certain results of Darroch and Morris (1967 , 1968) about the probability distributions of general type of transition counts and of linear function of the sojourn times $A_t(i)$, $i \in S$.

Let N_{jk} denote the number of direct transitions required for a first visit to state k from the initial state j and let $\tau(j, k)$ denote the first passage time from j to k. We can formally define N_{jk} and $\tau(j, k)$ as follows :

$$N_{jk} = \min \{n \mid n \in Z^+,\ X_0 = j,\ X_n = k\} \qquad (8)$$

and

$$\tau(j, k) = \inf \{t \mid t > 0,\ X(t) = k,\ X(0) = j\} . \qquad (9)$$

We obtain the first two moments of N_{jk} in section 4 and those of $\tau(j, k)$ in section 5, both under the assumption that the Markov process is irreducible.

Suppose that the finite Markov process is reducible and that its first $m < M$ states $1, \ldots, m$ constitute an ergodic class E and that the remaining $M-m$ states $m+1, \ldots, M$, constitute the class T of transient states. Recall that every state $j \varepsilon T$ leads to every state $k \varepsilon E$ but that no state in E leads to any state in T. In section 6, we obtain the first two moments of the number N_{jk} of visits to $k \varepsilon T$ when the initial state is $j \varepsilon T$, before the process is absorbed in E. It is then easy to obtain the first two moments of total number $N_j = \sum_{k\varepsilon T} N_{jk}$ of transitions within T before the eventual and the almost sure absorption of the process in E.

Define the integral

$$\nu_j = \int_0^{\infty} I[X(t) = j]\, dt, \qquad j \varepsilon T \tag{10}$$

pointwise and let

$$\nu = \sum_{j\varepsilon T} \nu_j \quad . \tag{11}$$

These are random variables denoting the total time spent in state $j \varepsilon T$ and in T respectively. The moments of ν_j and ν are calculated in section 7. The results of sections 4-7 are based on the work of Kemeney and Snell (1961, 1976). In section 8 we illustrate the results of these sections with examples.

Strictly speaking, it is necessary to demonstrate that the functions $N_t(j, k)$, $A_t(j)$, N_{jk}, $\tau(j, k)$ and ν_j are all r.v.s on $(\Omega, \mathbb{F}, P)$. It is easy to see that the transition counts $N(t)$, $N_t(j, k)$, N_{jk} and N_j are r.v.s. The assumption of separability of the process implies that $\tau(j,k)$ is also a r.v. In case of the stochastic integrals $A_t(j)$ and ν_j, which are defined pointwise, observe that for almost all $w \in \Omega$, the integrands are step functions with values in $\{0, 1\}$. Moreover, in case of ν_j, the integrand vanishe almost surely for all sufficiently large t, in view of the almost sure absorption of the process in the set E of ergodic states. It follows that $A_t(j)$ and ν_j are also non-negative r.v.s on $(\Omega, \mathbb{F}, P)$.

In sections 4-7 we shall need the following matrix notation in addition to that introduced earlier. If A is M-square matrix $((a_{jk}))$, A_{dg} denotes the diagonal matrix $\text{diag}(a_{11}, \ldots, a_{MM})$. An rxs matrix with all elements equal to one is denoted by E_{rs}. If A is any matrix $((a_{jk}))$, then A_{sq} is the matrix of same order with a_{jk} replaced by a_{jk}^2. It is well-known that if P is the one-step transition probability matrix of an irreducible aperiodic Markov chain, then $P^n \to A = E_{M1}\, \alpha^T$ as $n \to \infty$, with identical rows α^T, where $\alpha = (\alpha_1, \ldots, \alpha_M)^T$ is the unique positive solution of

$$\alpha^T P = \alpha^T \tag{12}$$

such that $\Sigma\, \alpha_j = 1$. If the chain is periodic (12) is no longer true. However, it is true that for every $\theta \in (0, 1)$

$$\lim_{n\to\infty} \{\theta\, I_M + (1 - \theta)\, P\}^n = A\,. \tag{13}$$

This result is sometimes refered to by saying that P is Euler - summable [cf. emeney and Snell (1976), p. 23].

2. MOMENTS OF TRANSITION COUNTS AND SOJOURN TIMES

Let $\{X(t), t \in \mathbb{R}^+\}$ be a finite Markov process with state-space $\{1, \ldots, M\}$. In what follows we shall write $q(i,j)$ for the intensity rate q_{ij}, $i \neq j$, and $q(i)$ for q_i as a notational convenience. The following theorem specifies the distribution of the random vector V_t defined by equation (1.5).

Theorem 1 : Let $q^*(i,j) = (1 - \delta_{ij})\, q(i, j)$ and let

$$p(i) = \Pr[X(0) = i]\ ,\ i \in S, \quad \sum_{i=1}^{m} p_i = 1.$$

Then

$$\Pr[N(t) = 0,\ X(0) = x_0] = p(x_0)\exp\{-q(x_0)t\}\ ,\ t \in \mathbb{R}^+,\ x_0 \in S. \tag{1}$$

For $n \geq 1$,

$$\Pr[N(t) = n,\ X_j = x_j\ \ j = 0,1,\ldots,n,\ T_j \leq t_j,\ j = 0,1,\ldots,n-1]$$

$$= p(x_0)\exp\{-q(x_n)t\}$$

$$\times \int_{S_n} \prod_{j=0}^{n-1} [q^*(x_j, x_{j+1}) \exp[-\{q(x_j) - q(x_n)\}u_j]du_j], \tag{2}$$

where $x_j \in S$ and

$$S_n = \{(u_0, u_1, \ldots, u_{n-1})\ |\ 0 \leq u_j \leq t_j,\ j = 0,\ldots,n-1,\ \sum_{j=0}^{n-1} u_j \leq t\}.$$

Proof : The relation (1) follows from theorem 2.6.2 and from the fact that

$$\Pr[N(t) = 0,\ X_0 = x_0] = \Pr[X(0) = x_0]\Pr[X(u) = x_0, 0<u\leq t\ |\ X(0) = x_0]$$

$$= p(x_0)\Pr[T_0 > t_0]$$

$$= p(x_0)\exp\{-\,q(x_0)t\}\ .$$

Let $n \geq 1$ and observe that the events A and B defined by

$$A = [N(t) = n] \cap \bigcap_{j=0}^{n} [X_j = x_j] \cap \bigcap_{j=0}^{n-1} [T_j \leq t_j] \tag{3}$$

$$B = \bigcap_{j=0}^{n} [X_j = x_j] \cap [\sum_{j=0}^{n} T_j \geq t] \cap S_n^* , \tag{4}$$

where $S_n^* = [(T_0, \ldots, T_{n-1}) \varepsilon S_n]$, are the same. It follows by virtue of theorem 2.6.3 and the Markov property that

$$\Pr[X_j = x_j \mid X_0 = x_0, \ldots, X_{j-1} = x_{j-1}, T_0, \ldots, T_{j-1}]$$

$$= q^*(x_{n-1}, x_n) / q(x_{n-1}) ; \tag{5}$$

$$\Pr[T_j \leq t_j \mid X_0 = x_0, \ldots, X_j = x_j, T_0, \ldots, T_{j-1}]$$

$$= 1 - \exp \{- q(x_j) t_j\} ; \tag{6}$$

and

$$\Pr[X_0 = x_0, T_0 \leq t_0] = p(x_0) [1 - \exp\{-q(x_0)t_0\}]. \tag{7}$$

The equations (5), (6) and (7) and the Markov property imply that for $n \geq 1$,

$$\Pr\{ \bigcap_{j=0}^{n} [X_j = x_j] \cap \bigcap_{j=0}^{n} [T_j \leq t_j] \}$$

$$= \prod_{j=0}^{n-1} [\{ q(x_j, x_{j+1})/q(x_j)\}[1 - \exp\{- q(x_j) t_j\}]]$$

$$\times [1 - \exp\{- q(x_n)t\}] p(x_0) . \tag{8}$$

Using the above results and the equality of the events A and B, one has

$$Pr(B) = p(x_0) \int_{S_n} \prod_{j=0}^{n-1} [q^*(x_j, x_{j+1}) \exp[-\{q(x_j)u_j\}] du_j]$$

$$\times \left[\int_{t-\sum_{j=0}^{n-1} t_j}^{\infty} q(x_n) \exp\{-q(x_n)u_n\} \, du_n \right]$$

$$= p(x_0)\exp\{-q(x_n)t\} \int_{S_n} \prod_{j=0}^{n-1} [q^*(x_j,x_{j+1})\exp[-\{q(x_j)-q(x_n)\}u_j]du_j],$$

which completes the proof.

Lemma 1 : Let $N(t)$ be the total number of jumps of the finite irreducible Markov process in $[0, t]$. Then there exist non-negative constants, α and β , $\beta > 0$, such that for all $h \in (0,1)$, $n \geq 1$,

$$Pr[N(h) \geq n] \leq \beta \int_0^h \{u^{n-1} e^{\alpha u}/(1-u)\}du . \tag{9}$$

Proof : Observe that for $n \geq 1$

$$Pr[N(h)=n]=\sum Pr\{[N(h)=n] \cap \bigcap_{j=0}^{n} [X_j=x_j] \cap \bigcap_{j=0}^{n-1} [T_j \leq h] \cap \sum_{j=0}^{n-1} T_j \leq h]\}, \tag{10}$$

where the summation extends over all $x_0, x_1, \ldots, x_n \in S$ such that $x_r \neq x_{r+1}$, $r = 0, 1, \ldots, (n-1)$. By theorem 1,a typical term in the sum on the right of (10) is

$$p(x_0)\exp\{-q(x_n)t\}\underset{0\le\sum u_j\le h}{\int_0^h\cdots\int_0^h}\prod_{j=0}^{n-1} q^*(x_j,x_{j+1})\exp[\{q(x_n)-q(x_j)\}u_j]du_j$$

$$\le \gamma^n \underset{0\le\sum u_j\le h}{\int_0^h\cdots\int_0^h} \exp\{\alpha\sum_{j=0}^{n-1} u\}\, du_0,\ldots,du_{n-1},$$

where

$$\gamma = \max\{q(i,j)\mid i,j\in S,\ i\ne j\}>0$$

and

$$\alpha = \max\{q(i)-q(j)\mid i,j\in S\}\ge 0 .$$

One can use induction to show that

$$\underset{0\le\sum u_j\le h}{\int_0^h\cdots\int_0^h} \exp\{\alpha\sum_{j=0}^{n-1} u_j\}du_0\cdots du_{n-1} = \frac{1}{(n-1)!}\int_0^h u^{n-1}e^{\alpha u}\,du .$$

The number of terms in the summation on the right side of (10) is $M(M-1)^n$. Pooling the above facts, one has

$$\Pr[N(h)=n]\le\{\gamma^n M(M-1)^n/(n-1)!\}\int_0^h u^{n-1}e^{\alpha u}\,du$$

$$\le \beta\int_0^h u^{n-1}e^{\alpha u}\,du ,$$

where $\beta = M\gamma(M-1)\exp\{\gamma(M-1)\}>0$. The inequality (9) follows immediately.

Corollary 1 : (i) $\Pr[N(h) \geq n] = o(h)$, $h \downarrow 0$,

(ii) $$\sum_{n=2}^{\infty} \Pr[N(h) \geq n] \leq \beta \int_0^h \{ue^{\alpha u}/(1-u)^2\}du = o(h), \quad h \downarrow 0.$$

Lemma 2 : If $i \neq j$, $i, j \in S$, then

$$\Pr[N_h(i, j) = 1] = p(i)\, q(i, j)\, h + o(h)$$

$$= E\, N_h(i, j) + o(h), \quad h \downarrow 0. \tag{11}$$

Proof : Observe that

$$\Pr[N_h(i,j) = 1] = \sum_{n=1}^{\infty} \sum_{r \in S} \Pr[N_h(i,j) = 1 \mid N(h) = n, X(0) = r]$$

$$\times \Pr[N(h) = n, X(0) = r].$$

$$= \Pr[N_h(i,j) = 1 \mid N(h) = 1, X(0) = i] \Pr[N(h) = 1, X(0) = i]$$

$$+ \sum_{r \neq i} \Pr[N_h(i,j) = 1 \mid N(h) = 1, X(0) = r] \Pr[N(h) = 1, X(0) = r]$$

$$+ \sum_{n \geq 2} \sum_{r \in S} \Pr[N_h(i,j) = 1 \mid N(h) = n, X(0) = r] \Pr[N(h) = n, X(0) = r]. \tag{12}$$

It is easy to see that for $r \neq i$

$$\Pr[N_h(i, j) = 1 \mid N(h) = 1, X(0) = r] \Pr[N(h) = 1, X(0) = r] = 0 \tag{13}$$

and that the third term in (12) is bounded above by

$$\sum_{n\geq 2} \sum_{r\varepsilon S} \Pr[N(h) = n, \quad X(0) = r]$$

$$\leq \sum_{n\geq 2} P[N(h) = n] = o(h) , \tag{14}$$

by (ii) of corollary (1). Moreover.

$$\Pr[N_h(i, j) = 1 \mid N(h) = 1, X(0) = i] \Pr[N(h) = 1, X(0) = i]$$

$$= p(i) \Pr[N(h) = 1|X(0) = i]\Pr[N_h(i,j) = 1|N(h) = 1, X(0) = i]$$

$$= p(i) \{q_i h + o(h)\} \{q_{ij}/q_i\} = p(i) q_{ij}h + o(h) .$$

The first part of (11) follows by virtue of (12), (13) and (14). To establish the second part of (11), one has

$$E[N_h(i, j)] = \sum_{n=1}^{\infty} \Pr[N_h(i, j) \geq n]$$

$$= \Pr[N_h(i,j) = 1] + \Pr[N_h(i,j) \geq 2] + \sum_{n\geq 2} \Pr[N_h(i,j) \geq n]$$

$$= p(i) q_{ij}h + o(h)$$

by virtue of the corollary 1 and the proof is complete.

We now introduce the two-dimensional Kronecker delta

$$\delta(u, v; x, y) = \begin{cases} 1 , & \text{if } u = x \text{ and } v = y, \\ 0 , & \text{otherwise} . \end{cases} \tag{15}$$

Theorem 2 : Let $p_i(t) = \Pr[X(t) = i], \quad i \varepsilon S.$ Then for $i \neq j$, $i, j \in S$

$$E[N_t(i, j)] = q(i, j) \int_0^t p_i(u)\, du, \tag{16}$$

$$E[N_t(i,j)\, N_t(r, s)] = \delta(i,j;\, r, s)\, q(i,j) \int_0^t p_i(u)\, du$$

$$+ q(i,j)\, q(r,s) \int_0^t \int_0^v \{p_{si}(v-u)\, p_r(u) + p_{jr}(v-u)\, p_i(u)\} du dv \,. \tag{17}$$

Proof : Let $\{[(k-1)h, kh), k = 1,\ldots, m\}$ be a partition of the interval $[0, t)$ and $n_k(i, j)$ denote the number of direct transitions from state i to state j, $i \neq j$, during the interval $[(k-1)h, kh)$, $k = 1, \ldots, m$, where $h = t/m$. It is easy to see that

$$N_t(i, j) = \sum_{k=1}^{n} n_k(i, j) \,.$$

If m is so large that $h < 1$, corollaries (1) and (2) of lemma 1 imply that

$$\Pr[n_k(i, j) \geq 2\,] = o(h)$$

and $\sum_{n\geq 2} \Pr[n_k(i, j) \geq n] = o(h)$, $h \downarrow 0$.

Observe further that for $\ell \neq j$ and $r \neq i$,

$$\Pr[n_k(i,j) = 1 \;, X(kh) = \ell \,, \; X((k-1)h) = r]$$

$$= \{q(r,i)h + o(h)\} \{q(i,j)h + o(h)\} \{q(j,\ell)h + o(h)\} + o(h)$$

$$= \; o(h) \,.$$

It follows, therefore, that

$$\Pr[n_k(i,j) = 1] = \sum_{\ell \in S} \sum_{r \in S} \Pr[n_k(i,j) = 1,\ X(\overline{k-1}\ h) = r,\ X(kh) = \ell]$$

$$= \Pr[n_k(i, j) = 1,\ X(\overline{k-1}\ h) = i,\ X(kh) = j] + o(h)$$

$$= \Pr[X(\overline{k-1}\ h) = i]\ \{q(i, j)h + o(h)\} + o(h)$$

$$= P_i(\overline{k-1}\ h)\ q(i, j)h + o(h).$$

Thus

$$E[N_t(i, j)] = \sum_{k=1}^{m} p_i(\overline{k-1}\ h)\ q(i, j)h + o(h)\ , \qquad (18)$$

which is a Riemann sum approximating the integral $q(i, j) \int_0^t p_i(u)\,du$. This establishes (16) by allowing $h \to 0$ in the right of (18).

In order to obtain (17), observe that

$$E[N_t(i, j)\ N_t(r, s)] = \sum_{k=1}^{m} \sum_{\ell=1}^{m} E\{n_k(i,j)\ n_\ell(r, s)\}$$

where $n_k(i, j)$ and $n\ (r, s)$ are defined as before. It is easy to verify that

$$E[n_k(i, j)\ n_\ell(r,s)] = \Pr[n_k(i,j) = 1,\ n_\ell(r,s) = 1] + o(h)$$

and therefore we find that

$$\begin{aligned}
E[N_t(i, j)\, N_t(r,s)] = {} & \sum_{k=1}^{m} \Pr[n_k(i,j) = 1,\ n_\ell(r,s) = 1] \\
& + \sum_{k=2}^{m} \Pr[n_k(i,j) = 1,\ n_{k-1}(r, s) = 1] \\
& + \sum_{k=2}^{m} \Pr[n_{k-1}(i,j) = 1,\ n_k(r, s) = 1] \\
& + \sum_{k=3}^{m} \sum_{\ell=1}^{k-2} \Pr[n_k(i, j) = 1,\ n_\ell(r, s) = 1] \\
& + \sum_{\ell=3}^{m} \sum_{k=1}^{\ell-2} \Pr[n_k(i, j) = 1,\ n_\ell(r,s) = 1] + o(h) \\
= {} & \sum_{k=1}^{m} q(i,j)h\, p_i(\overline{k-1}\, h)\ \ \delta(i, j;\, r, s) \\
& + \sum_{k=2}^{m} q(i,j)\, q(j,s)\, h^2\, p_r(\overline{k-2}\, h) \\
& + \sum_{k=2}^{m} q(i,j)\, q(j,s) h^2\, p_i(\overline{k-2}\, h) \\
& + \sum_{k=3}^{m} \sum_{\ell=1}^{k-2} q(i,j)\, q(r,s)\, h^2 p_{si}(\overline{k-1-\ell} h)\, p_r(\overline{\ell-1}\, h) \\
& + \sum_{\ell=3}^{m} \sum_{k=1}^{\ell-2} q(i,j)\, q(r,s) h^2\, p_{jr}(\overline{\ell-1-k}\, h)\, p_i(\overline{k-1}\, h) + o(h).
\end{aligned} \tag{19}$$

Now letting $m \to \infty$ in the above we get (17). [cf. Apostol(1975), p. 400], after identifying the Riemann integrals whose approximating sums appear on the right of (19).

We now turn to moments of $A_t(i)$, $i \in S$.

<u>Theorem 3</u> : The first two moments of $A_t(i)$ are given by

$$E[A_t(i)] = \int_0^t p_i(u)\,du, \tag{20}$$

$$E[A_t(i)A_t(j)] = \int_0^t \int_0^u \{p_{ji}(u-v)p_j(v) + p_{ij}(u-v)p_i(u)\}dv\,du. \tag{21}$$

<u>Proof</u> : The result (20) is an immediate consequence of Fubini's theorem and the equation (1.7) for $A_t(i)$. Using Fubini's theorem, we have

$$E\{A_t(i)\, A_t(j)\} = E\{\int_0^t \int_0^t I[X(u) = i]\, I[X(v) = j]du\,dv$$

$$= \int_0^t \int_0^t E[I\{[X(u)=i] \cap [X(v) = j]\}]du\,dv$$

$$= \int_0^t \int_0^t \Pr[X(u) = i, X(v) = j]\, du\,dv$$

$$= \int_0^t \int_0^u \Pr[X(v) = j, X(u) = i]\, dv\,du$$

$$+ \int_0^t \int_0^v \Pr[X(u) = i, X(v) = j]\, du\,dv$$

$$= \int_0^t \int_0^v P_j(v)p_{ji}(u-v)dv\,du + \int_0^t \int_0^v p_i(u)p_{ij}(v-u)du\,dv,$$

which is (21).

The technique used above can be used to establish the following theorem which we state without proof.

Theorem 4 : Let $i, j, r, s \in S$. Then

$$E[N_t(i,j)\, A_t(r)]$$

$$= q(i,j) \int_0^t \int_0^u \{p_{ri}(u-v)\, p_r(v) + p_{jr}(u-v)\, p_i(v)\} dv\, du \qquad (22)$$

and

$$E[N_t(i,j) - q(i,j)\, A_t(i)][N_t(r,s) - q(r,s)\, A_t(r)]$$

$$= \delta(i, j; r, s) \int_0^t p_i(u)\, du. \qquad (23)$$

Suppose now that the finite Markov process is irreducible and that its intensity matrix has simple characteristic roots $\alpha_0=0$, $\alpha_2, .., \alpha_M$. Then, if $(A)_{jk}$ denotes the (j,k)-element of A,

$$p_k(t) = Pr[X(t) = k] = \sum_{j=1}^{M} p_j(0)\, p_{jk}(t)$$

$$= \pi_k + \sum_{j=1}^{M} p_j(0) \sum_{r=2}^{M} \exp\{\alpha_r t\}(\xi_r\, \eta_r^T)_{jk} ,$$

where $\{\pi_1, \ldots, \pi_M\}$ is the invariant distribution of the process and the notation of section 2.4 is used. Recall that $Re(\alpha_r) < 0$ by lemma 2.4.2. It is then easy to establish the following

Corollary 2 : If the finite Markov process is irreducible and its intensity matrix has simple characteristic roots then, as $t \to \infty$,

(i) $E[N_t(i,j)/t] \to \pi_i\, q(i,j)$

(ii) $E[N_t(i,j)\ N_t(r,s)/t^2] \to \pi_i \pi_r\ q(i,j)\ q(r,s)$

(iii) $E[A_t(i)/t] \to \pi_i$

(iv) $E[A_t(i)\ A_t(j)/t^2] \to \pi_i \pi_j$

(v) $E[N_t(i,j)\ A_t(r)/t^2] \to \pi_i \pi_r\ q(i,j)$

(vi) $E[\{N_t(i,j) - q(i,j)\ A_t(i)\}\{N_t(r,s) - q(r,s)\ A_t(r)\}/t]$

$$\to \delta(i,\ j\ ;\ r,\ s)\ \pi_i\ , \quad i,\ j,\ r,\ s \in S.$$

Example 1 : Recalling the expressions for the transition probabilities $p_{ij}(t)$, $i, j = 0, 1$ of the two-state Markov process of example 1.3.3, one can compute the first two moments of $N_t(0,1)$, $N_t(1,0)$, $A_t(0)$ and $A_t(1) = t - A_t(0)$. We state the following results only for the first moments of $N_t(0,1)$ and $A_t(0)$:

$$E[N_t(0,1)] = \frac{\lambda\mu t}{(\lambda+\mu)} + \frac{\lambda\ \{1 - e^{-(\lambda+\mu)t}\}}{(\lambda+\mu)^2}\ \{\lambda\ p_0(0) - \mu\ p_1(0)\}$$

$$E[A_t(0)] = \frac{\mu t}{\lambda+\mu} + \frac{\{1 - e^{-(\lambda+\mu)t}\}}{(\lambda+\mu)^2}\ \{\lambda\ p_0(0) - \mu\ p_1(0)\}\ .$$

3. PROBABILITY DISTRIBUTIONS ASSOCIATED WITH TRANSITION COUNTS AND SOJOURN TIMES

In this section we obtain the probability generating function (p.g.f.) of the number of transitions of a more general type to be described in the next paragraph. We also obtain the moment generating function (m.g.f.) of a linear combination of the sojourn times in different states. These results are based on the work of Darroch and Morris (1967, 1968).

Observe that for a Markov process with M states, there are at most M(M-1) possible direct transitions. Let A be a subset of the set of all possible direct transitions. An A-transition is said to occur at epoch t if some transition $(i \to j) \in A$ occurs at t. The following are some of the examples of such subsets A which are of interest.

Example 1 : Let A be the singleton set $\{(i \to j)\}$ consisting of only one transition of the process from a specified state i to a specified state j. The number of A-transitions in $(0, t]$ is then precisely the r.v. $N_t(i,j)$ whose moments were studied in section 2.

Example 2 : Let H be a subset of S. The set $A = \{(i \to j) | i \in S, j \in H\}$ consists of those transitions which lead to an occurrence of an H-state i.e., a state belonging to the set H. The set $A = \{(i \to j) \mid i \in S-H, j \in H\}$ consists of those transitions which lead to a state in H from outside H. If $H = \{j\}$, the number of A - transitions is the number of visits to j.

Let $N_t(A)$ denote the number of A-transitions in the interval $(0, t]$. We obtain the p.g.f. of $N_t(A)$ in the following

Theorem 1 : Let Q be the intensity matrix and A be a subset of the set of all possible transitions $(i \to j)$. Define

$$\Delta(i, j, A) = \begin{cases} 1, & \text{if } (i \to j) \in A, \\ 0, & \text{otherwise;} \end{cases}$$

$Q(A) = ((q(i,j)\ \Delta(i,j,A)))$ and $Q(A^c) = Q - Q(A)$. The p.g.f. of $N_t(A)$ is given by

$$E\{\xi^{N_t(A)}\} = p^T(0) \exp\{tQ(A,\xi)\}\, E_{M1}, \tag{1}$$

where $Q(A,\xi) = Q(A^c) + \xi Q(A)$ and $|\xi| \leq 1$.

Proof : Define $\mu_j(\xi,t) = E\{\xi^{N_t(A)}\ I[X(t) = j]\}$, $j \in S$ and let $\mu(\xi,t)$ be the Mx1 column vector $(\mu_1(\xi,t),\ldots,\mu_M(\xi,t))^T$. Observe that the probability of two or more jumps of the process in the interval (t, t+h] being o(h), we can assert that, for h > 0

$$N_{t+h}(A) = N_t(A) + \Delta(X(t), X(t+h), A)$$

with probability 1 - o(h). Hence

$$\begin{aligned}
&\mu_j(\xi, t+h)\\
&= E\{\xi^{N_{t+h}(A)}\ I[X(t+h) = j]\\
&= E[E\{\xi^{N_t(A)}\ \xi^{\Delta(X(t),\, j,A)}\ I[X(t+h) = j] \mid X(t),\ N_t(A)\}] + o(h)\\
&= \sum_{i\in S} p_{ij}(h)\ E\{\xi^{N_t(A)}\ \xi^{\Delta(i,j,A)}\ I[X(t) = i]\} + o(h)\\
&= \sum_{i\in S} \{q_{ij}h + o(h)\}\ \xi^{\Delta(i,j,A)}\ \mu_i(\xi, t)\} + o(h)
\end{aligned}$$

Hence, in matrix notation

$$\mu^T(\xi, t+h) = \mu^T(\xi, t)\ \{I_M + h\ Q(A,\xi)\} + o(h) .$$

Subtracting $\mu^T(\xi, t)$ from both sides, dividing by h and allowing $h \downarrow 0$, one has the differential equation

$$d\mu^T(\xi,t)/dt = \mu^T(\xi, t)\, Q(A,\xi) .$$

Its solution, subject to the obvious initial conditions $\mu(\xi,0) = p(0)$, is [cf. section 2.3]

$$\mu^T(\xi, t) = p^T(0) \exp\{tQ(A,\xi)\} . \tag{2}$$

Since $E[\xi^{N_t(A)}] = \sum_{i \epsilon S} E\{\xi^{N_t(A)} I[X(t) = i]\}$,

we have (1) as a consequence of (2).

Example 1 : Consider the two state Markov process described in example 1.3.3. Here the set of all possible transitions is $\{(0 \to 1), (1 \to 0)\}$. Define $A = \{(0 \to 1)\}$. Then

$$Q(A) = \begin{pmatrix} 0 & \lambda \\ 0 & 0 \end{pmatrix} , \qquad Q(A^c) = \begin{pmatrix} -\lambda & 0 \\ \mu & -\mu \end{pmatrix} ;$$

so that the p.g.f. of $N_t(0,1)$ is

$$E[\xi^{N_t(0,1)}] = [p_0(0), p_1(0)] \exp\left(t \begin{pmatrix} -\lambda & \lambda \\ \mu & -\mu \end{pmatrix}\right) \begin{pmatrix} 1 \\ 1 \end{pmatrix} .$$

Similarly,

$$E[\xi^{N_t(1,0)}] = [p_0(0), p_1(0)] \exp\left(t \begin{pmatrix} -\lambda & \lambda \\ \mu & -\mu \end{pmatrix}\right) \begin{pmatrix} 1 \\ 1 \end{pmatrix}$$

and

$$E[\xi^{N_t}] = [p_0(0),\ p_1(0)] \exp\left(t \begin{pmatrix} -\lambda & \xi\lambda \\ \xi\mu & -\mu \end{pmatrix} \right) \begin{pmatrix} 1 \\ 1 \end{pmatrix} .$$

Example 2 : In the case of random walk with reflecting barriers, discussed in example 2.5.1, the number $R(t)$ of jumps to the right and the number $L(t)$ of jumps to the left are of interest. In case of $R(t)$, $A = \{(i, i+1) \mid i = 0, 1, \ldots, M-1\}$ and for $L(t)$, we should take $A = \{(i, i-1) \mid i = 1, \ldots, M\}$. Thus for $R(t)$

$$Q_{R(t)}(A, \xi) = ((\ q^*(i, j)\ )) ,$$

where the non-zero values of $q^*(i,j)$ are

$$q^*(i, i+1) = \xi\lambda , \quad i = 0, 1, \ldots, M-1$$

$$q^*(i, i) = q(i, i), i = 0, 1, \ldots, M ,$$

$$q^*(i, i-1) = q(i, i-1), \ i = 1, \ldots, M.$$

These assertions can be used to obtain the p.g.f. of $N_t(A)$.

Let $m \le M$, $d_1, \ldots, d_m$ be non-negative constants and $A_t(i)$ denote the sojourn time of the Markov process in state i during the time interval $[0, t)$. Define

$$V(t) = \sum_{i=1}^{m} d_i A_t(i) , \quad t \in \mathbb{R}^+ . \tag{3}$$

That such linear functions are of interest becomes obvious if we consider, for example, the machine-repairmen problem of Example 3.7.4 and take $d_j = j$, so that $V(t) = \sum_{j=1}^{M} j\, A_t(j)$. This represents the total machine time lost during $[0, t)$. In the following theorem, we obtain the m.g.f. of $V(t)$.

<u>Theorem 2</u> : Let D be the mxm matrix $((\delta_{ij}\, d_j))$, $i,j = 1,\ldots,m$ and let D^* be the MxM matrix

$$D^* = \begin{pmatrix} D & 0 \\ 0 & 0 \end{pmatrix} .$$

The m.g.f. of $V(t)$ is given by

$$E\{\exp(-\theta V(t))\} = p^T(0) \exp\{t(Q - \theta D^*)\} E_{M1} , \qquad (4)$$

where $\theta \in \mathbb{R}^+$.

<u>Proof</u> : Let $h > 0$ and define

$$W[X(t), X(t+h)]h = V(t+h) - V(t) = \sum_{j\varepsilon S} d_j \{A_{t+h}(j) - A_t(j)\} ,$$

$$\varphi_j(\theta, t) = E\{e^{-\theta V(t)} I[X(t) = j]\}, \quad j \varepsilon S ,$$

and

$$\varphi(\theta, t) = (\varphi_1(\theta, t), \ldots, \varphi_M(\theta, t))^T .$$

As in theorem 1, consider

$$\varphi_j(t+h) = E[\exp\{-\theta V(t+h)\} I[X(t+h) = j]]$$

$$= E[E\{e^{-\theta V(t)} e^{-\theta h W(X(t),X(t+h))} I[X(t+h)=j] | X(t),V(t)\}]$$

$$= \sum_{i\varepsilon S} p_{ij}(h) E\{e^{-\theta V(t)} I[X(t)=i]\} \exp\{-\theta h\, W(i,j)\}.$$

Expanding $\exp\{-\theta h\, W(i, j)\}$ and using the fact that $p_{ij}(h) = q_{ij}h + o(h)$, after some simplification, one has

$$\varphi_j(t+h) = \sum_{i\varepsilon S} E\{e^{-\theta V(t)}\, I[X(t) = i] g_{ij}\} + o(h) ,$$

where $g_{ij} = \delta_{ij} + h\{q_{ij} - \delta_{ij}\ \theta W(i, j)\}$. Now, with probability $1 - o(h)$, there is no change of state in $(t, t+h]$ so that with probability $1 - o(h)$,

$$W(i, i) = \begin{cases} d_i & \text{if } i \leq m, \\ 0 & \text{if } i > m . \end{cases}$$

Let G be the MxM matrix $((g_{ij}))$, so that in matrix notation, one has

$$\varphi^T(\theta, t+h) = \varphi^T(\theta, t)\ [I_M + h(Q - \theta D^*) + o(h)] .$$

The usual techniques lead to the differential equation

$$d\varphi^T(\theta, t)/dt = \varphi^T(\theta, t)\ [Q - \theta D^*] .$$

Solving this equation with the initial condition $\varphi(\theta,0) = p(0)$, one has

$$\varphi^T(\theta, t) = p^T(0) \exp[t(Q - \theta D^*)]$$

and therefore, (4) follows from the fact that

$$E[e^{-\theta V(t)}] = \varphi^T(\theta, t) E_{M1} .$$

The proof is complete.

Example 3 : Consider the machine-repairman problem described in example 3.7.4 and suppose $M = 2$, $r = 1$. Let

$$V(t) = \sum_{j=0}^{2} j\,A_t(j) = A_t(1) + 2\,A_t(2).$$

Here

$$Q = \begin{pmatrix} -2\lambda & 2\lambda & 0 \\ \mu & -(\lambda+\mu) & \lambda \\ 0 & \mu & -\mu \end{pmatrix}, \quad D^* = \text{diag}\,\{0,1,2\}\,.$$

Thus the m.g.f. of $V(t)$ is

$$(p_0(0), p_1(0), p_2(0))\,\exp\left\langle t \begin{pmatrix} -2\lambda & 2\lambda & 0 \\ \mu & -(\lambda+\mu+\theta) & \mu \\ 0 & \mu & -\mu-2\theta \end{pmatrix} \right\rangle \begin{pmatrix} 1 \\ 1 \\ 1 \end{pmatrix}.$$

4. MOMENTS OF THE FIRST PASSAGE TRANSITION COUNTS : IRREDUCIBLE MARKOV PROCESS

Let $\{X(t),\ t \in \mathbb{R}^+\}$ be a finite, irreducible Markov process on the state-space $S = \{1, \ldots, M\}$. In this section we obtain the first two moments of N_{jk} defined by equation (1.8). The r.v. N_{jk} is defined in terms of the Markov chain $\mathbf{X}(R)$ only. Its one-step transition probabilities are

$$r_{jk} = \{q_{jk}/q_j\}\ \{1 - \delta_{jk}\}\ , \quad j,\ k \in S,$$

where all $q_j > 0$ because of the assumption of irreducibility of the process. Thus all states of the $\mathbf{X}(R)$ chain are persistent, non-null but could be periodic as illustrated in section 3.3. In what follows let $h_{jk} = E\,N_{jk}$ and $g_{jk} = E\,N_{jk}^2$ with $H = ((h_{jk}))$ and $G = ((g_{jk}))$ as the corresponding M-square matrices.

Lemma 1 : All moments of N_{jk} are finite.

Proof : Consider a new Markov chain on $S = \{1, \ldots, M\}$ in which state k is converted into an absorbing state. The transition probabilities of the new chain are

$$r^*_{uv} = \begin{cases} r_{uv} & \text{if } u \neq j,\ k \neq v, \\ 1 & \text{if } u = j,\ v = k \\ 0 & \text{if } u = j,\ v \neq k,\ u,v = 1,\ldots,M. \end{cases}$$

Since the transitions before absorption in state k are governed by the original transition probabilities, N_{jk} represents the number of steps to absorption in state k. Since $j \to k$, there exists a sequence $j_1, \ldots, j_n$ of distinct states, $j \neq j_1 \neq \ldots \neq j_{n-1} \neq k$, such that

$$r_{jj_1}\, r_{j_1 j_2} \cdots r_{j_{n-1}k} > 0 .$$

It is obvious that $n \leq M-1$. Therefore, there exists a positive number $\alpha(j)$ such that

$$\Pr[N_{jk} \leq M] \geq \alpha(j)\ , \qquad 0 < \alpha(j) \leq 1 .$$

If $\alpha(j) = 1$, the statement of the lemma is true. So let $\alpha(j) < 1$. Define $p = \min\{\alpha(j) \mid j \neq k,\ j \in S\}$ which, by definition, is positive and less than one. One can now easily see that

$$\Pr[N_{jk} > M] \leq (1 - p)$$

and therefore

$$\Pr[N_{jk} > nM] \leq (1 - p)^n .$$

More generally, one finds that

$$\Pr[N_{jk} > n] \leq (1 - p)^{[n/M]}$$

where $[n/M]$ is the integral part of n/M. In other words, there exist positive constants α and β, $0 < \beta < 1$, such that

$$\Pr[N_{jk} > n] \leq \alpha\beta^n , \quad n = 1, 2, \ldots \tag{1}$$

It follows that $\sum_n n^r \Pr[N_{jk} > n] < \infty$ for all $r \in Z^+$. Hence all moments of N_{jk} are finite [cf. Feller (1969), p. 148].

<u>Lemma 2</u> : The matrix H satisfies the equation

$$H = R\{H - H_{dg}\} + E_{MM} . \tag{2}$$

<u>Proof</u> : Observe that for all $j, k \in S$

$$h_{jk} = E[E\{N_{jk} \mid X_1\}]$$

$$= \sum_{i \neq k} r_{ji} \{h_{ik} + 1\} + r_{jk}$$

$$= \sum_{i=1}^{M} (r_{ji} h_{ik}) - r_{jk} h_{kk} + 1,$$

of which (2) is an immediate consequence.

Lemma 3 : If $\alpha = (\alpha_1, \ldots, \alpha_M)^T$ is the unique invariants probability vector of R, then

$$h_{jj} = 1/\alpha_j, \qquad j = 1, \ldots, M . \tag{3}$$

Proof : By definition $\alpha^T R = \alpha^T$ so that pre-multiplication of (2) by α^T yields

$$\alpha^T H = \alpha^T H - \alpha^T H_{dg} + \alpha^T E_{MM} ,$$

which implies $\alpha^T H_{dg} = \alpha^T E_{MM}$ and hence (3) follows.

Lemma 4 : The equation (2) satisfied by H has a unique solution.

Proof : The proof is by contradiction. Let, if possible, H and H^* be two solutions of equation (2). By Lemma 3,

$$h_{jj} = h^*_{jj} = 1/\alpha_j, \quad j = 1, \ldots, M.$$

Therefore,

$$H - H^* = R[H - H_{dg}] - R[H^* - H^*_{dg}]$$

$$= R(H - H^*) ,$$

i.e., every column of $H - H^*$ is a right characteristic vector of R corresponding to the characteristic root $\lambda = 1$. In view of the equation (1.13), we have for every $\theta \in (0, 1)$, as $n \to \infty$,

$$\{\theta I_M + (1 - \theta) R\}^n \to A ,$$

where $A = E_{M1}\alpha^T$. One now easily has

$$\{\theta I_M + (1 - \theta) R\}^n (H - H^*) = (H - H^*)$$

for all $n \geq 1$ from which it follows that

$$E_{M1}\,\alpha^T(H - H^*) = H - H^* . \tag{4}$$

Equating (i,j)-elements of matrices on both sides of this equation we obtain

$$h_{ij} - h^*_{ij} = \sum_{s=1}^{M} \alpha_s\{h_{sj} - h^*_{sj}\}$$

for each fixed i and for all j = 1, ..., M. It follows that $h_{ij} - h^*_{ij}$ is independent of i. Since this difference vanishes for i = j, the lemma is established.

<u>Lemma 5</u> : Let $R(\theta) = \theta I_M + (1 - \theta)R$, $0 < \theta < 1$, α be the invariant probability vector of R and $A = E_{M1}\alpha^T$. Then

(i) $RA = AR = A$,

(ii) $Z = [I_M - (R - A)]^{-1}$ exists,

(iii) $RZ = ZR$, $ZE_{M1} = E_{M1}$,

(iv) $\alpha^T Z = \alpha^T$,

(v) $(I_M - R)Z = (I_M - A)$.

<u>Proof</u> : The result (i) is easily obtained by using the facts that $A = E_{M1}\alpha^T$ and $\alpha^T R = \alpha^T$.

In order to obtain (ii) observe that by (i)

$$(R - A)^n = R^n - A , \qquad n \geq 1, \tag{5}$$

so that for $\theta \in (0,1)$,

$$\sum_{k=0}^{n} \binom{n}{k} \theta^{n-k} (1-\theta)^k (R-A)^k$$

$$= \{\theta\ I_M + (1-\theta)\ R\}^n - A \to 0, \qquad (6)$$

as $n \to \infty$. Define $B = R - A$ and consider

$$(I_M - B) \sum_{k=0}^{n} \binom{n}{k} \theta^{n-k} (1-\theta)^k \{I_M + B + \ldots + B^k\}$$

$$= \sum_{k=0}^{n} \binom{n}{k} \theta^{n-k} (1-\theta)^k \{I_M - B^{k+1}\}$$

$$= I_M - \{\sum_{k=0}^{n} \binom{n}{k} \theta^{n-k} (1-\theta)^k B^k\}B$$

$$\to I_M$$

as $n \to \infty$ by virtue of (6). Thus $I_M - B$ has an inverse which is given by

$$\lim_{n\to\infty} \sum_{k=0}^{n} \binom{n}{k} \theta^{n-k} (1-\theta)^k \{I_M + B + \ldots + B^k\} \qquad (7)$$

which can be shown to be independent of θ.

The remaining results can be easily obtained by matrix manipulations employing (5) and (7).

Theorem 1 : The unique solution of equation (2) is

$$H = \{I_M - Z + E_{MM}\ Z_{dg}\}A_{dg}^{-1}. \qquad (8)$$

Proof : By virtue of lemma 4, it is enough to show that H as specified by (8) satisfies (2). One has

$$R(H - A_{dg}^{-1}) = \{- RZ + RE_{MM} Z_{dg}\} A_{dg}^{-1}$$

$$= H + \{Z - RZ - I_M\} A_{dg}^{-1}$$

$$= H - AA_{dg}^{-1} \quad , \quad \text{by (iv) and (v) of lemma 5,}$$

$$= H - E_{MM} \; .$$

Further, by lemma 3, $A_{dg}^{-1} = H_{dg}$, so that

$$H = R(H - H_{dg}) + E_{MM}$$

which is (2) and (8) is established.

Lemma 6 : The matrix G is a solution of the equation

$$G = R\{G - G_{dg}\} - 2R\{Z - E_{MM} Z_{dg}\}A_{dg}^{-1} + E_{MM} \; . \tag{9}$$

Proof : Observe that

$$g_{jk} = \sum_{i \neq k} r_{ji} \; E(N_{ik} + 1)^2 + r_{jk}$$

$$= \sum_{i \neq k} r_{ji} g_{ik} + 2 \sum_{i \neq k} r_{ji} h_{ik} + 1 \; ,$$

so that

$$G = R(G - G_{dg}) + 2R(H - H_{dg}) + E_{MM}.$$

The equation (9) is obtained by substituting for $(H - H_{dg})$ from equation (8).

Lemma 7 : The diagonal entries g_{jj} of G are given by

$$G_{dg} = A_{dg}^{-1} \{2 Z_{dg} A_{dg}^{-1} - I_M \}. \tag{10}$$

Proof : Premultiply equation (9) by α^T to obtain

$$\alpha^T G = \alpha^T\{G - G_{dg}\} - 2\alpha^T\{Z - E_{MM} Z_{dg}\} A_{dg}^{-1} + E_{1M} .$$

Using result (iv) of lemma 5 and the fact that $\alpha^T A_{dg}^{-1} = E_{1M}$, one obtains

$$\alpha^T G_{dg} = 2E_{1M} Z_{dg} A_{dg}^{-1} - E_{1M} ,$$

or equivalently

$$g_{jj} = 2 z_{jj}/\alpha_j^2 - 1/\alpha_j, \qquad j = 1, \ldots, M$$

of which (10) is the matrix representation.

Theorem 2 : The equation (9) has a unique solution which is specified by

$$G = H\{2Z_{dg} A_{dg}^{-1} - I_M\} + 2\{ZH - E_{MM} (ZH)_{dg}\} . \tag{11}$$

Proof : The uniqueness of the solution of (9) follows as in lemma 4 and (11) follows by a direct verification that it satisfies (9).

Remark : The matrix Z which is essential for evaluating H and G, is known as the fundamental matrix of the Markov chain $\mathbf{X}(R)$. We had to employ Euler-summability arguments in obtaining Z because of the possibility that $\mathbf{X}(R)$ is a periodic chain. If a finite chain is an irreducible ergodic chain with transition matrix P and invariant probability vector α, then it is easy to see that

$$Z = I_M - (P - \Pi)^{-1} \tag{12}$$

exists, where $\Pi = \lim P^n = E_{M1}\alpha^T$ and that

$$Z = I_M + \sum_{n=1}^{\infty} (P^n - \Pi) , \tag{13}$$

has all the properties listed in lemma 5. In fact the results of theorems 1 and 2 remain valid for such an irreducible, aperiodic finite Markov chain, with Z as defined by (12) or (13).

5. MEAN AND VARIANCE OF FIRST PASSAGE TIMES

An irreducible finite Markov process passes through all its states with probability one. Hence the r.v. $\tau(j,k)$ defined by equation (1.9) and which is a continuous-time analogue of N_{jk} is of considerable interest. Let $\tau(j,j)$ denote the duration between two successive returns of the process to state j. If ξ_j denotes the sojourn time in state j, then

$$\tau(j, j) = \xi_j + \tau\{X(\xi_j), j\} . \tag{1}$$

We obtain the first two moments of $\tau(j,k)$, $j, k \in S$ for an irreducible finite Markov process with state-space $S = \{1, \ldots, M\}$.

Recall that if $\mathbf{X} = \{X(t), t \in \mathbb{R}^+\}$ is such a finite Markov process, every discrete skeleton $\mathbf{X}(h) = \{X(nh, n \in Z^+\}$, to scale $h > 0$, is an irreducible, aperiodic and therefore ergodic Markov chain with one-step transition probability matrix $P(h)$. Let $\tau_h(j,k)$, $j \neq k$, denote the number N_{jk} of transitions required for a first passage from j to k in the $\mathbf{X}(h)$-chain. It is easy to see that

$$0 \leq h\,\tau_h(j,k) - \tau(j,k) \leq h$$

with probability one and therefore $h\,\tau_h(j,k) \to \tau(j,k)$ almost surely as $h \downarrow 0$. Thus by the Fatou-Lebesgue theorem [cf. Loève (1968) p. 125]

$$\lim_{h\to 0} E[h\tau_h(j,k)] = E[\tau(j,k)] = u_{jk}, \text{ say;} \tag{2}$$

$$\lim_{h\to 0} E[h^2\,\tau_h^2(j,k)] = E[\tau^2(j,k)] = v_{jk}\,, \text{ say;} \tag{3}$$

and by virtue of lemma 4.1, u_{jk} and v_{jk} are both finite.

Let $u_{jk}(h) = E\{\tau_h(j,k)\}$, $v_{jk}(h) = E\{\tau_h^2(j,k)\}$ $j \neq k$, and $U(h)$ and $V(h)$ denote the corresponding matrices $((u_{jk}(h)))$ and $((v_{jk}(h)))$. Define,

$$U^*(h) = U(h) - [U(h)]_{dg}\,,$$

$$V^*(h) = V(h) - [V(h)]_{dg}\,.$$

Then by theorems 4.1 and 4.2

$$U^*(h) = \{E_{MM}\,Z(h))_{dg} - Z(h)\}\,\Pi_{dg}^{-1} \tag{4}$$

and

$$V^*(h) = U^*(h)\,\{2(Z(h))_{dg}\,\Pi_{dg}^{-1} - I_M\}$$

$$+\ 2\,Z(h)\,U(h) - 2E_{MM}(Z(h)\,U(h))_{dg}\,, \tag{5}$$

where

(i) $\Pi = \lim_{t\to\infty} P(t) = E_{M1}(\pi_1, \ldots, \pi_M)$,

(ii) $(\pi_1, \ldots, \pi_M) = (\pi_1, \ldots, \pi_M)\, P(t)$ for all $t > 0$,

is the invariant probability vector corresponding to the $P(t)$-matrix,

(iii) $Z(h) = (I_M - P(h) + \Pi)^{-1}$ is the fundamental matrix of the $\mathbf{X}(h)$-chain.

In order to obtain $U^* = \lim h\, U^*(h)$ and $V^* = \lim h^2\, V^*(h)$, we need the following lemmas.

Lemma 1 : The matrix

$$W = \int_0^\infty \{P(t) - \Pi\}dt = \left(\left(\int_0^\infty \{p_{jk}(t) - \pi_k\}dt \right)\right)$$

has finite elements.

Proof : This is a straightforward consequence of the geometric rate of convergence of $P(t)$ to its limit Π as $t \to \infty$. [cf. Theorem 3.5.1].

Lemma 2 : The matrix $(\Pi - Q)$ is non-singular and $(\Pi - Q)^{-1} = \Pi + W$.

Proof : It is easy to observe that $Q\Pi = 0$ and that

$$\Pi P(t) = P(t)\Pi = \Pi = \Pi^2 .$$

Thus

$$(\Pi - Q)[\{\Pi + \int_0^t \{P(u) - \Pi\}du]$$

$$= \Pi - \int_0^t QP(u)du$$

$$= \Pi - \int_0^t \{dP(u)/du\}\, du$$

$$= \Pi - P(t) + I_M \to I_M$$

as $t \to \infty$. This establishes the assertions of the lemma.

Theorem 1 : The matrix

$$U^* = \{E_{MM} W_{dg} - W\} \Pi_{dg}^{-1} . \tag{6}$$

Proof : Observe that, as $h \to 0$,

$$\begin{aligned} \lim h Z(h) &= \lim h\{I_M - P(h) + \Pi\}^{-1} \\ &= \lim h[I_M + \sum_{n=1}^{\infty} \{P(nh) - \Pi\}] \\ &= \int_0^{\infty} \{P(t) - \Pi\}dt = W. \end{aligned} \tag{7}$$

Similarly, $h E_{MM}(Z(h))_{dg} \to E_{MM} W_{dg}$ as $h \downarrow 0$. Hence (6) follows by virtue of equations (2) and (4).

Theorem 2 : The matrix

$$V^* = 2\{U^* W_{dg} \Pi_{dg}^{-1} + WU^* - E_{MM}(WU^*)_{dg}\} . \tag{8}$$

Proof : By virtue of equations (5) and (7) and theorem 1, we have

$$\begin{aligned} \lim_{h \to 0} h^2 V^*(h) &= \lim h^2 [U^*(h) \{2(Z(h))_{dg} \Pi_{dg}^{-1} - I_M\} \\ &\quad + 2 Z(h) U(h) - 2 E_{MM}(Z(h) U(h))_{dg}] \\ &= 2U^* W_{dg} \Pi_{dg}^{-1} + W U^* - E_{MM}(WU)_{dg} . \end{aligned}$$

In the above, we may replace the matrix U in the second and third terms by U^*, if we recall that $u_{jj}(h) = 1/\pi_j$ and so $h u_{jj}(h) \to 0$. Thus finally we have (8).

Let u_{jj} and v_{jj} denote the first two moments of the duration $\tau(j, j)$ between two successive returns of the process to state j and let ξ_j denote the sojourn time of the process in state j. At the end of the sojourn time ξ_j, the process shifts to a state k with probability q_{jk}/q_j and then returns for the first time to state j after duration $\tau(k,j)$. Recall that ξ_j has exponential distribution with mean $1/q_j$ and one can verify that ξ_j and τ_{kj} are independent r.v.s. Consequently,

$$u_{jj} = E\{\tau(j, j)\} = 1/q_j + \sum_{k\neq j} (q_{jk}/q_j)\, u_{kj} \tag{9}$$

and $v_{jj} = E\,\tau^2(j, j)$

$$= 2/q_j^2 + \sum_{k\neq j} (q_{jk}/q_j)\, v_{kj} + (2/q_j)\Sigma(q_{jk}/q_j)u_{kj}\,. \tag{10}$$

<u>Theorem 3</u> : We have

$$u_{jj} = 1/\{\pi_j q_j\} \quad , \quad j = 1, \ldots, M. \tag{11}$$

<u>Proof</u> : Let $U_{dg} = \mathrm{diag}\,(u_{11}, \ldots, u_{MM})$. Then by (9)

$$\begin{aligned} U_{dg} &= -Q_{dg}^{-1} - \{Q_{dg}^{-1}\, Q\, U^*\}_{dg} \\ &= -Q_{dg}^{-1} - [Q_{dg}^{-1}\, Q\{E_{MM}\, W_{dg} - W\}\Pi_{dg}^{-1}]_{dg} \\ &= -Q_{dg}^{-1} + [Q_{dg}^{-1}\, Q\, W\, \Pi_{dg}^{-1}]_{dg} \end{aligned}$$

since $Q\,E_{MM} = 0$. In order to evaluate the second term, observe that

$$\begin{aligned} Q\,W &= Q \int_0^\infty \{P(t) - \Pi\}dt \\ &= \int_0^\infty Q\,P(t)\,dt, \\ &= \int_0^\infty (dP(t)/dt)\,dt \\ &= \Pi - I_M\,. \end{aligned} \tag{12}$$

Thus

$$\begin{aligned} U_{dg} &= -Q_{dg}^{-1} + [Q_{dg}^{-1}\;\Pi\;\Pi_{dg}^{-1}]_{dg} - Q_{dg}^{-1}\;\Pi_{dg}^{-1} \\ &= -Q_{dg}^{-1}\;\Pi_{dg}^{-1}\;, \end{aligned}$$

which is the matrix representation of (11).

<u>Theorem 4</u> : We have

$$v_{jj} = 2u_{jj}^2\,\{q_j w_{jj} + \pi_j\},\quad j = 1, \ldots, M. \tag{13}$$

<u>Proof</u> : Let $V_{dg} = \mathrm{diag}\,(v_{11}, \ldots, v_{MM})$. Then by equation (10),

$$V_{dg} = 2Q_{dg}^{-2} - [Q_{dg}^{-1}\;Q\,V^* - 2Q_{dg}^{-2}\;QU^*]_{dg} \tag{14}$$

and from equation (8)

$$\begin{aligned} QV^* &= 2Q\{U^*\;W_{dg}\;\Pi_{dg}^{-1} + WU^* - E_{MM}(WU^*)_{dg}\} \\ &= 2QU^*\;W_{dg}\;\Pi_{dg}^{-1} + 2QWU^*\,. \end{aligned} \tag{15}$$

But equations (6) and (12) imply that

$$QU^* = Q\{E_{MM} W_{dg} \Pi_{dg}^{-1} - W\Pi_{dg}^{-1}\}$$

$$= - QW\Pi_{dg}^{-1}$$

$$= (I_M - \Pi)\Pi_{dg}^{-1} \qquad (16)$$

Further $QW U^* = (\Pi - I_M) U^*$

$$= (\Pi - I_M) E_{MM} W_{dg} - W \Pi_{dg}^{-1}$$

$$= (W - \Pi W) \Pi_{dg}^{-1}$$

$$= W\Pi_{dg}^{-1} \quad , \qquad (17)$$

since $\Pi E_{MM} = E_{MM}$ and

$$\Pi W = \int_0^\infty \{\Pi P(t) - \Pi^2\}dt = \int_0^\infty (\Pi - \Pi)dt = 0 \ .$$

Using (15), (16) and (17) in (14), one has

$$V_{dg} = 2Q_{dg}^{-1} - 2[Q_{dg}^{-1}(I_M - \Pi)\Pi_{dg}^{-1} W_{dg} \Pi_{dg}^{-1} + Q_{dg}^{-1} W\Pi_{dg}^{-1}$$

$$- Q_{dg}^{-2}(I_M - \Pi)\Pi_{dg}^{-1}]_{dg} \ . \qquad (18)$$

A routine computation now yields (13).

6. TRANSITION COUNTS FOR AN ABSORBING MARKOV PROCESS

Let $\{X(t), t \in \mathbb{R}^+\}$ be a finite Markov process with state space $S = \{1, 2, \ldots, M\}$ such that the subset $E = \{1, 2, \ldots, m\}$, $1 \leq m < M$, is an irreducible class of ergodic states and its complement $T = \{m+1, \ldots, M\}$ is the class of transient states. The transition probability matrix of such a process is representable in the following partitioned form :

$$P(t) = \begin{pmatrix} A(t) & 0 \\ C(t) & B(t) \end{pmatrix} ,$$

where $A(t)$ is an mxm stochastic matrix, $B(t)$ an $(M-m) \times (M-m)$ sub-stochastic matrix and $C(t)$ is $(M-m) \times m$ matrix. The corresponding partitioned form of the intensity matrix Q is

$$Q = \begin{pmatrix} A & 0 \\ C & B \end{pmatrix} .$$

It is obvious that every state $j \varepsilon T$ leads to a state $k \varepsilon E$ and that no state in E leads to any state in T. Thus $\sum_{k \varepsilon T} p_{jk}(t)$ represents the probability that the Markov process starting in a transient state j remains in T during $[0, t]$. The probability that it remains forever in T is, therefore,

$$\lim_{t \to \infty} \sum_{k \varepsilon T} p_{jk}(t) = 0 ,$$

by theorem 3.5.1. In other words, a finite Markov process with initial state $j \varepsilon T$ eventually reaches the ergodic class E with probability one. We, therefore, seek the first two moments of the total number N_{jk} of visits to state $k \varepsilon T$ when the initial state is $j \varepsilon T$. We also calculate the first two moments of the number $N_j = \sum_{k \varepsilon T} N_{jk}$ of changes of state in T before absorption in E, when the initial state is $j \varepsilon T$. As in section 4, these quantities can be obtained in terms of the jump matrix R which can now be represented in the partitioned form

$$R = \begin{pmatrix} R_{11} & 0 \\ R_{21} & R_{22} \end{pmatrix} .$$

Lemma 1 : Let $d_{ij} = E(N_{ij})$, $i, j = m+1, \ldots, M$, and let $D = ((d_{ij}))$. Then

$$D = (I_{M-m} - R_{22})^{-1} . \tag{1}$$

Proof : It is easy to see that

$$(I_{M-m} - R_{22}) \sum_{r=0}^{n} R_{22}^{r} = I_{M-m} - R_{22}^{n} \to I_{M-m}$$

as $n \to \infty$. Thus $(I_{M-m} - R_{22})$ is non-singular and its inverse is given by

$$(I_{M-m} - R_{22})^{-1} = \sum_{n=0}^{\infty} R_{22}^{n} .$$

Now observe that

$$d_{ij} = \delta_{ij} + \sum_{k \in T} (q_{ik}/q_i) d_{kj} , \quad i, j \in T;$$

so that in matrix notation

$$D = I_{M-m} + R_{22} D$$

of which (1) is an immediate consequence.

Remark : The matrix $(I_{M-m} - R_{22})^{-1}$ is called the fundamental matrix of the **X**(R) chain and therefore we shall reserve the symbol Z for it.

Lemma 2 : Let $f_{ij} = E\, N_{ij}^2$, $i, j \varepsilon T$. Then

$$F = ((f_{ij})) = Z\{2Z_{dg} - I_{M-m}\}. \qquad (2)$$

Proof : Use the usual conditional expectation argument to obtain

$$f_{ij} = \sum_{k\varepsilon E} (q_{ik}/q_i)\delta_{ij} + \sum_{k\varepsilon T} (q_{ik}/q_i)\, E\{N_{kj} + \delta_{ij}\}^2$$

$$= \delta_{ij} + \sum_{k\varepsilon T} (q_{ik}/q_i)\, E\, N_{kj}^2 + 2\delta_{ij} \sum_{k\varepsilon T} (q_{ik}/q_i)\, E(N_{kj}) ,$$

which in matrix notation is

$$F = I_{M-M} + R_{22} F + 2(R_{22} Z)_{dg} .$$

Hence,

$$F = Z\{I_{M-m} + 2(R_{22} Z)_{dg}\} . \qquad (3)$$

But $R_{22}Z = R_{22} \sum_{n=0}^{\infty} R_{22}^n = Z - I_{M-m}$, of which (2) is an immediate consequence by virtue of (3).

Lemma 3 : If $N^* = (N_{(m+1)}, \ldots, N_M)^T$, then

$$E(N^*) = Z\, E_{(M-m)1} \qquad (4)$$

and

$$E(N_{sq}^*) = (2Z - I_{M-m})\, Z\, E_{(M-m)1} . \qquad (5)$$

Proof : Since $N_i = \sum_{j\varepsilon T} N_{ij}$, (4) is a consequence of lemma 1.

To compute $E(N_{sq}^*)$, observe that for $i = m+1, \ldots, M$,

$$E(N_i^2) = \sum_{k\varepsilon E} q_{ik}/q_i + \sum_{k\varepsilon T} (q_{ik}/q_i)\, E(N_k+1)^2$$

$$= \sum_{k\varepsilon T} (q_{ik}/q_i)\, E(N_k^2) + 2 \sum_{k\varepsilon T} (q_{ik}/q_i)\, E(N_k) + 1,$$

which in matrix notation becomes

$$E(N^*_{sq}) = R_{22}\, E(N_{sq}) + 2\, R_{22}\, Z\, E_{M1} + E_{M1}.$$

The equation (5) follows immediately as in lemma 2.

7. SOJOURN TIMES FOR ABSORBING MARKOV PROCESSES

Let $\{X(t),\ t \varepsilon \mathrm{IR}^+\}$ be a finite absorbing Markov process as described in section 6. The moments of the r.v.s ν_j and ν defined by equations (1.10) and (1.11) are calculated in this section. Define for $h > 0$ and $i \varepsilon T$,

$$\nu_i(h) = \sum_{n=0}^{\infty} I[X(nh) = i]\ , \quad \nu(h) = \sum_{i\varepsilon T} \nu_i(h)\ , \tag{1}$$

which are r.v.s denoting the total number of visits to state i and the total time spent in T by the discrete skeleton $\mathbf{X}(h) = \{X(nh),\ n \varepsilon Z^+\}$. It is easy to verify that

$$0 \leq h\nu(h) - \nu \leq h \tag{2}$$

so that with probability one

$$\lim_{h\downarrow 0} h\nu\,(h) = \nu\ . \tag{3}$$

Moreover, since $h\, \nu_i(h)$ is a Riemann sum approximating the stochastic Riemann integral ν_i defined by equation (1.10),

$$\lim_{h \downarrow 0} h\, \nu_i(h) = \nu_i \tag{4}$$

with probability one.

Let

$$J_i^{(r)} = E[\nu^r \mid X(0) = i]\ , \quad K_{ij}^{(r)} = E[\nu_j^r \mid X(0) = i]\ ,$$

$$J_i^{(r)}(h) = E[\{\nu(h)\}^r \mid X(0) = i]\ , \tag{5}$$

$$K_{ij}^{(r)}(h) = E[(\nu_j(h))^r \mid X(0) = i]\ , \qquad i,\ j = m+1,\dots,M,\ r \geq 1$$

and let $J^{(r)}$, $J^{(r)}(h)$ and $K^{(r)}$, $K^{(r)}(h)$ denote the corresponding $(M-m) \times 1$ vectors and $(M-m)$ - square matrices.

Lemma 1 : With the notation of section 6, the $(M-m)$-square matrix B of intensity rates corresponding to the transient states is non-singular and

$$B^{-1} = -\int_0^\infty B(t)\, dt. \tag{6}$$

The matrix $-B^{-1}$ is called the fundamental matrix of the process and we shall denote it by N.

Proof : Observe that for all $i, j \in T$, $p_{ij}(t) \to 0$ as $t \to \infty$ and by virtue of theorem 3.5.1, there exist positive numbers a and b, $0 < b < 1$, such that

$$p_{ij}(t) \leq ab^t\ , \quad t \in \mathbb{R}^+\ .$$

It follows that for all $i, j \in T$

$$\int_0^\infty p_{ij}(t)\,dt < \infty .$$

The forward differential equations (2.2.8) for $p_{ij}(t)$, i, j ε T, are, in matrix notation

$$dB(t)/dt = B\,B(t)$$

and therefore

$$B \int_0^\infty B(t)dt = \int_0^\infty [dB(t)/dt]dt = I_{M-m}$$

from which (6) follows.

Lemma 2 : For all i, j ε T and $r \geq 1$, the moments $J_i^{(r)}$ and $K_{ij}^{(r)}$ are finite.

Proof : By definition $\nu_j \leq \nu$ and therefore it is enough to show that $J_i^{(r)} < \infty$. Observe then that $\nu > t$ iff $X(t) \varepsilon T$. Thus

$$P[\nu \leq t \mid X(0) = i] = 1 - \sum_{j\varepsilon T} p_{ij}(t), \quad 0 \leq t < \infty.$$

It follows that

$$J_i^{(r)} = E[\nu^r \mid X(0) = i] = \int_0^\infty t^r d\{1 - \sum_{j\varepsilon T} p_{ij}(t)\}$$

$$= -\sum_{j\varepsilon T} \int_0^\infty t^r\{dp_{ij}(t)/dt\}dt$$

$$= -\sum_{j\varepsilon T} \sum_{k\varepsilon T} q_{ik} \int_0^\infty t^r p_{ki}(t)\,dt$$

$$\leq \sum_{j\varepsilon T} \{q_i - \sum_{k\varepsilon T-\{i\}} q_{ik}\} \int_0^\infty t^r ab^t\,dt <$$

where a and b are as defined in lemma 1. The proof is complete.

Lemma 3 : The moments $J_i^{(r)}$, $J_i^{(r)}(h)$ and $K_{ij}^{(r)}$ and $K_{ij}^{(r)}(h)$ are related by the equations

$$\begin{aligned} J_i^{(r)} &= \lim_{h \downarrow 0} h^r \, J_i^{(r)}(h) \\ K_{ij}^{(r)} &= \lim_{h \downarrow 0} h^r \, K_{ij}^{(r)}(h) \quad , \quad i, j \in T. \end{aligned} \tag{7}$$

Proof : These results are simple consequences of relations (2), (3), (4), lemma 2 and Fatou-Lebesgue theorem [cf. Loẹve (1968), p. 125].

Lemma 4 : The matrix $K^{(1)}$ and the vector $J^{(1)}$ are given by

$$K^{(1)} = N, \qquad J^{(1)} = N\, E_{(M-m)1} \quad . \tag{8}$$

Proof : We have

$$\begin{aligned} E[\nu_j \mid X(0) = i] &= E\{ \int_0^\infty I[X(t) = j]dt \mid X(0) = i \} \\ &= \int_0^\infty E\{I[X(t) = j] \mid X(0) = i\}dt \\ &= \int_0^\infty p_{ij}(t)dt \end{aligned}$$

by the Fubini theorem [cf. Loẹve (1968), p. 136]. The expression for $K^{(1)}$ is a consequence of lemma 1 and that for $J^{(1)}$ follows by definition of ν .

<u>Theorem 1</u> : For $r \geq 1$,

$$J^{(r)} = r!\ N^r E_{(M-m)1} \tag{9}$$

and

$$K^{(r)} = r!\ N(N_{dg})^{r-1} . \tag{10}$$

<u>Proof</u> : In what follows, let E_i denote the conditional expectation conditional on $X(0) = i$. Standard techniques can be easily used to see that for $i \in T$,

$$E_i\{h\ \nu(h)\}^r = \sum_{\ell \in E} \{h^r p_{i\ell}(h) + \sum_{\ell \in T} p_{i\ell}(h)\ E_\ell[h^r\{\nu(h) + 1\}^r] \tag{11}$$

$$= \sum_{\ell \in E} h^r p_{i\ell}(h) + \sum_{\ell \in T} p_{i\ell}(h) \sum_{t=0}^{r} \binom{r}{t} h^{r-t} E_\ell\{h\nu(h)\}^t$$

or equivalently

$$E_i\{h\ \nu(h)\}^r - \sum_{\ell \in T} p_{i\ell}(h)\ E_\ell\ \{h\ \nu(h)\}^r$$

$$= h^r + \sum_{\ell \in T} p_{i\ell}(h) \sum_{t=1}^{r-1} \binom{r}{t} h^{r-t}\ E_\ell\{h\ \nu(h)\}^t .$$

In matrix notation, one has

$$\{I_{M-m} - B(h)\}\ J^{(r)}(h) = h^r\ E_{(M-m)1} + \sum_{t=1}^{r-1} \binom{r}{t} h^{r-t}\ J^{(t)}(h) .$$

Divide both sides by h and allow $h \downarrow 0$ to obtain

$$-B\ J^{(r)} = r\ J^{(r-1)} , \qquad r \geq 2 ,$$

or equivalently $J^{(r)} = r\ N\ J^{(r-1)}$ of which (10) is a simple consequence.

The argument used to derive equation (11) can also be used to see that for $r > 1$,

$$E_i\{h\ \nu_j(h)\}^r$$

$$= \delta_{ij}\ h^r \sum_{\ell \varepsilon E} p_{i\ell}(h) + \sum_{\ell \varepsilon T} p_{i\ell}(h)\ E_\ell[h^r\{\nu_j(h) + \delta_{ij}\}^r]$$

$$= \delta_{ij}\ h^r \sum_{\ell \varepsilon E} p_{i\ell}(h) + \sum_{\ell \varepsilon T} p_{i\ell}(h) \sum_{t=0}^{r-1} \binom{r}{t} h^{r-t} E_\ell\{h^t \nu_j^t(h)\} \delta_{ij}^{r-t}$$

$$= \delta_{ij}^r\ h^r + \sum_{\ell \varepsilon T} p_{i\ell}(h)\ E_\ell\{h^r\ \nu_j^r(h)\}$$

$$+ \sum_{\ell \varepsilon T} p_{i\ell}(h) \sum_{t=1}^{r-1} \binom{r}{t} h^{r-t}\ E_\ell\{h^t\ \nu_j^t(h)\}\ \delta_{ij}\ .$$

It is now easy to see that for all $i,\ j\ \varepsilon\ T$

$$[\{1 - p_{ii}(h)\}/h]\ E_i\{h\ \nu_j(h)\}^r \sum_{\ell \varepsilon T-\{i\}} \{ p_{i\ell}(h)/h\}\ E_\ell\{h\nu_j(h)\}^r$$

$$= \sum_{\ell \varepsilon T} p_{i\ell}(h) \sum_{t=1}^{r-1} \binom{r}{t} h^{r-t-1}\ E_\ell\{h^t\ \nu_j^t(h)\}\ \delta_{ij}\ .$$

Allowing $h \downarrow 0$ in the above equation one has

$$-\ q_{ii} E_i(\nu_j^r) + \sum_{\ell \varepsilon T-\{i\}} q_{i\ell}\ E_\ell\{\nu_j^r\}$$

$$= r\ E_i\{\ \nu_j^{r-1}\}\ \delta_{ij}\ ,\ i,\ \ j\ \varepsilon\ T.$$

Employing matrix notation, it follows that

$$-\ B\ K^{(r)} = r\{K^{(r-1)}\}_{dg}$$

or equivalently,

$$K^{(r)} = r\, N\{K^{(r-1)}\}_{dg} , \qquad r \geq 2 . \tag{12}$$

The result (10) is now an easy consequence of (12) and (8).

8. EXAMPLES

The theory developed in sections 4-7 is illustrated in this section with two examples.

Example 1 : A Markov process in Radio Biology

Reid (1953) describes a Markov model in the study of injury and recovery in biological systems following irradiation. The occurrence of an effective event in a sensitive volume of an organism is considered as a transition of that volume to a new state. The model is formulated as a birth-death process on the finite state-space $\{1, \ldots, M\}$ with

$$q_{ii+1} = \lambda_i, \quad i = 1,\ldots, M-1, \quad q_{ii-1} = \mu_i , \quad i = 2,\ldots,M, \tag{1}$$

as the only non-zero intensity rates. We shall discuss the simple case obtaining when $\lambda_i \equiv \lambda$, $\mu_i \equiv \mu$, $\lambda \neq \mu$. In this case the Markov process is a random walk with reflecting barriers on the state-space $S = \{1, \ldots, M\}$. [cf. example 2.5.1] . This is an irreducible ergodic Markov process and its invariant distribution is specified by

$$\pi_j = (1 - \rho)\rho^{j-1}/\{1 - \rho^M\} , \quad j = 1,\ldots, M , \tag{2}$$

where $\rho = (\lambda/\mu)$.

The non-zero transition probabilities of the associated $\mathbf{X}(R)$- chain are

$$r_{12} = r_{M(M-1)} = 1, \quad r_{i(i+1)} = \lambda/(\lambda+\mu) , \quad i = 2,\ldots,M-1$$

$$r_{ii-1} = \mu/(\lambda+\mu) , \quad i = 2, \ldots, M.$$

The Markov chain is an irreducible, persistent, non-null chain which is periodic with period 2. The invariant probability vector $\alpha = (\alpha_1, \ldots, \alpha_M)^T$, for the $\mathbf{X}(R)$-chain is the unique solution of the equations

$$\alpha^T = \alpha^T R , \qquad E_{1M}\alpha = 1.$$

In fact, one has

$$\begin{aligned} \alpha_1 &= (1 - \rho)/\{2(1 - \rho^{M-1})\} , \\ \alpha_j &= (1 - \rho^2)^{j-1}/\{2(1 - \rho^{M-1})\} , \quad j = 2,\ldots, M-1, \qquad (3) \\ \alpha_M &= (1 - \rho)\rho^{M-2}/\{2(1 - \rho^{M-1})\} . \end{aligned}$$

If $A = E_{M1}\alpha^T$, the associated fundamental matrix is $Z = (I-R+A)^{-1}$ specified by equation (4.7).

Observe that

$$Z = \lim_{n\to\infty} \sum_{k=0}^{n} \binom{n}{k} \theta^{n-k} (1-\theta)^k \{I + R + \ldots + R^k - kA\}$$

to evaluate which it is enough to obtain the spectral resolution of R. By adopting the method of example 2.5.1, one can demonstrate that R has the following distinct characteristic roots:

$$\beta_0 = 1,\quad \beta_r = 2\sqrt{pq}\cos\{\pi r/(M-1)\}\ ,\ r = 1,\ldots,M-2,\ \beta_{M-1} = -1,$$

where $p = (\lambda/(\lambda+\mu)$ and $q = \mu/(\lambda+\mu)$. The left characteristic vector $\eta_r = (y_1(r), \ldots, y_M(r))^T$ corresponding to the root β_r is α if $r = 0$ and is specified as follows for $r = 1,\ldots,M-2$.

$$y_j(r) = \rho^{j/2}\ \{\sin \frac{\pi rj}{M-1} - \rho^{-1} \sin \frac{\pi r(j-2)}{M-1}\}\ ,\ j = 1,\ldots, M.$$

If $r = M-1$,

$$y_1(M-1) = 1,\quad y_j(M-1) = (-1)^{j+1}\, q^{-1}\, \rho^{j-2},\quad j = 2,\ldots, M-1$$

$$y_M(M-1) = (-1)^{M-1}\ \rho^{M-2}\ .$$

The right characteristic vector $\xi_r = (x_1(r), \ldots, x_M(r))^T$ corresponding to the root β_0 is E_{M1} and for $r = M-1$, $x_j(M-1) = (-1)^{j+1}$, $1 = 1,\ldots, M$. Furthermore

$$x_j(r) = \rho^{-j/2}\ \{\sin \frac{\pi rj}{M-1} - \rho^{-1} \sin \frac{\pi r(j-2)}{M-1}\}$$

for $j = 1,\ldots, M$, $r = 1, \ldots, M-2$. The normalizing constants $c(r)$ which ensure that $c(r)\xi_r^T\eta_r = 1$, are

$$c(0) = 1,\quad c(M-1) = (1-\rho)/\{2(1-\rho^{M-1})\}\ ,$$

$$c(r) = [\sum_{j=1}^{M} \{\sin(\frac{\pi rj}{M-1}) - \rho^{-1} \sin(\frac{\pi r(j-2)}{M-1})\}^2]^{-1}\ .$$

Let $A_0 = A$, $A_r = \xi_r\eta_r^T$, $r = 1, \ldots, M-1$. Using the fact that $\sum_{r=0}^{M-1} A_r = I_M$, one has

$$I + R + \ldots + R^k - kA = A + \sum_{r=1}^{M-1} (1 - \alpha_r)$$

$$= A + \sum_{r=1}^{M-1} \frac{(1 - \beta_r^k)}{1 - \beta_r} A_r$$

from which one can easily conclude that

$$Z = A + \sum_{r=1}^{M-1} (1 - \beta_r)^{-1} A_r .$$

Thus equations (4.8) and (4.11) can be used to write down explicit expressions for h_{ij} and g_{ij} defined in section 4.

In order to write down the mean u_{jk} and the second moment v_{jk} of the first passage time $\tau(j,k)$, it is enough to obtain the matrix W

$$W = \int_0^\infty \{P(t) - \Pi\}dt$$

for our Markov process. Observe that the spectral representation of $P(t)$ is

$$P(t) = \Pi + \sum_{r=1}^{M-1} \exp(\beta_r t) A_r ,$$

where $\beta_0 = 0$ and $\beta_r = -(\lambda+\mu) + 2\sqrt{\lambda\mu} \cos(\pi r/M)$, $r = 1,\ldots,M-1$, are the M distinct characteristic roots of Q and $A_r = \xi_r \eta_r^T$, ξ_r and η_r being the right and left characteristic vectors of Q corresponding to the root β_r; [cf. Example 2.5.1]. Since $\beta_r < 0$ for $r = 1,\ldots, M-1$, it follows that

$$W = \sum_{r=1}^{M-1} \beta_r^{-1} A_r .$$

It is now not difficult to write down explicit expressions for u_{jk} and v_{jk} by using equations 5.6, 5.8, 5.9 and 5.10 in conju-ction with example 2.5.1.

Example 2 : Fix and Neyman model for treatment of cancer.

As already indicated in example 1.1.2, Fix and Neyman (1951) suggested a Markov model for the study of treatment of a disease like cancer. This model has 4 states S_1, S_2, S_3 and S_4, where a patient is in state S_1 when he is diagonsed as a patient suffering from the disease. He is in state S_2 if he dies from the disease and is in state S_3 if he has recovered. The state S_4 represents his death from some other cause or withdrawal from the study. It is obvious that states S_2 and S_4 are absorbing states which are accessible from S_1 and S_3 which in turn communicate amongst themselves. Thus the only non-zero intensity rates are q_{12}, q_{13} and q_{31} and q_{34}. Relabelling the states S_1, S_2, S_3 and S_4 as 3, 1, 4 and 2 respectively, the new intensity matrix is

$$\begin{pmatrix} 0 & 0 & 0 & 0 \\ 0 & 0 & 0 & 0 \\ q_{12} & 0 & q_{11} & q_{13} \\ 0 & q_{34} & q_{31} & q_{33} \end{pmatrix} . \qquad (4)$$

Here $E = \{1, 2\}$ is the set of ergodic states and $T = \{3, 4\}$ is the set of transient states.

The intensity matrix can be expressed in the partitioned form

$$Q = \begin{pmatrix} A & 0 \\ C & B \end{pmatrix}$$

where

$$A = \begin{pmatrix} 0 & 0 \\ 0 & 0 \end{pmatrix}, \quad B = \begin{pmatrix} q_{11} & q_{13} \\ q_{31} & q_{33} \end{pmatrix} \quad \text{and} \quad C = \begin{pmatrix} q_{12} & 0 \\ 0 & q_{34} \end{pmatrix}.$$

The partitoned form of the corresponding jump matrix is

$$R = \begin{pmatrix} R_{11} & 0 \\ R_{21} & R_{22} \end{pmatrix}$$

with $R_{11} = I_2$,

$$R_{21} = \begin{pmatrix} r_{12} & 0 \\ 0 & r_{34} \end{pmatrix} \quad \text{and} \quad R_{22} = \begin{pmatrix} 0 & r_{13} \\ r_{31} & 0 \end{pmatrix}.$$

Here

$$r_{12} = q_{12}/(q_{12} + q_{13}), \qquad r_{34} = q_{34}/(q_{34} + q_{31})$$

$$r_{13} = q_{13}/(q_{12} + q_{13}), \qquad r_{31} = q_{31}/(q_{34} + q_{31}).$$

The fundamental matrix D of the $\mathbf{X}(\mathbb{R})$ -chain is by definition

$$D = (I_2 - R_{22})^{-1}$$

$$= \frac{1}{\Delta}\begin{pmatrix} (q_{12} + q_{13})(q_{31} + q_{34}) & q_{13}(q_{31} + q_{34}) \\ q_{31}(q_{12} + q_{13}) & (q_{12} + q_{13})(q_{31}+q_{34}) \end{pmatrix} \qquad (5)$$

and the findamental matrix N of section 6 is

$$N = - B^{-1} = - \begin{pmatrix} q_{11} & q_{13} \\ q_{31} & q_{33} \end{pmatrix}^{-1}$$

$$= \begin{pmatrix} q_{12} + q_{13} & - q_{13} \\ - q_{31} & q_{34} + q_{31} \end{pmatrix}^{-1}$$

$$= \frac{1}{\Delta}\begin{pmatrix} q_{34} + q_{31} & q_{13} \\ q_{31} & q_{12} + q_{13} \end{pmatrix} . \qquad (6)$$

Here and in (5)

$$\Delta = q_{12}\, q_{34} + q_{12}\, q_{31} + q_{13}\, q_{34} .$$

The following are some of the possible conclusions. If a patient has recovered from the disease, then the expected number of times he is attacked by it again, before his eventual death or withdrawal is $q_{31}\,(q_{12} + q_{13})/\Delta$ [cf. lemma 6.1] . A patient who has been diagnosed as suffering from the disease dies becuase of that disease with probability $q_{12}(q_{31} + q_{34})/\Delta$. Finally, the expected duration of survival of the patient before he dies or is lost from the study is $(q_{13} + q_{34} + q_{31})/\Delta$ [cf. lemma 7.4] .

CHAPTER V

STATISTICAL INFERENCE

1. INTRODUCTION

We have studied the analytic, probabilistic and statistical properties of a finite Markov process in the earlier chapters. Suppose we develop a finite Markov process as a model for a real life situation. It will be very rare that we shall be able to specify the 'values' of the intensity rates or of the parameters which define the intensity rates. Thus, for example, it would be unrealistic to assume that we know a priori, the values of the parameters λ and μ governing the machine-repairmen problem [cf. Example 3.7.4] or those governing the formation of elephant herds [cf. Example 2.5.3]. It is, therefore, reasonable to face the problem of estimation of the intensity rates or the parameters defining them, and of testing the hypotheses about such parameters on the basis of observations on the process.

In the literature two types of observational procedures are discussed. One either observes a single realization $\{X(u), 0 \leq u \leq t\}$ completely over the fixed interval [0, t] of time or observe $n > 1$ independent and identical copies of the process, each over the fixed duration [0, t]. We shall soon discover that it is almost impossible to obtain precise statistical properties of the usual estimators and to devise tests of hypotheses for a finite period of observation or for a finite number n of independent copies of the process. Thus the attention of the research workers in this area has been mainly concentrated on obtaining the asymptotic properties of the estimators and on devising test-statistic whose asymptotic distribution can be obtained.

Corresponding to the two procedures of observation mentioned above, there are two methods of studying the asymptotic properties of the different procedures of statistical inference. In case a single realization of the process is observed, the asymptotic properties are obtained as the period t of observation becomes large. In the second situation, t is kept fixed and the number n of observed realizations is allowed to grow to $+\infty$. It is easy to see that in this second situation, we do not face any major new problems as the classical theory of statistical inference from i.i.d. random variables is applicable. We shall therefore concentrate on studying the asymptotic properties for a single realization of a finite Markov process as the period t of observation tends to infinity.

One may feel that the requirement that a process be completely observable over $[0, t]$ is too unrealistic. There is some substance in this criticism. However, when it is not possible to observe the process completely over $[0, t]$, it should be possible to observe it, at, say, a lattice set $\{0, h, 2h, \ldots\}$ of time points. Thus one has a realization from the discrete skeleton $\{X(nh), n \in Z^+\}$ which is in fact a Markov chain [cf. definition 2.2.2]. We do not propose to discuss the problems of statistical inference for a Markov chain. The interested reader may refer to Billingsley (1961) and (1961a) for an introduction to this topic.

The procedure of statistical inference that we shall be

describing in the subsequent sections are all likelihood based. We therefore obtain the likelihood function in section 2. The maximum likelihood (ML) estimators of the intensity rates are obtained in section 3 and their asymptotic properties are studied in section 4. The next section, section 5, discusses the strong consistency of the c-approximate ML estimators of the parameters defining the intensity rates. It is well-known that ML estimators may not always exist. We therefore concentrate on studying the properties of the so-called ML equation estimators in section 6. The asymptotic distributions of the ML equation estimators are obtained in section 7. The usual likelihood ratio tests and the asymptotic distributions of the appropriate Neyman-Pearson statistics are described in section 8. Our concluding remarks are made in the last section, section 9.

2. THE LIKELIHOOD FUNCTION

Let $\{X(t),\ t \in \mathbb{R}^+\}$ be a finite Markov process on a complete probability space $(\Omega, \mathbb{F}, P)$. Let $S = \{1, \ldots, M\}$ be its state-space and $Q = ((q(i,j)))$ be its intensity matrix. We have already seen in section 4.1, that the complete observation $\{X(u),\ 0 \leq u \leq t\}$ of the Markov process on $[0, t]$ is equivalent to an observation on the random vector

$$V_t = \{N(t), X_0, T_0, \ldots, X_{N(t)-1}, T_{N(t)-1}, X_{N(t)}\}$$

of random dimension $2N(t)+1$. Here $N(t)$ is the number of changes of state during $[0, t]$, $X_0, \ldots, X_{N(t)}$ represent the states visited and $T_0, \ldots, T_{N(t)-1}$ are the sojourn times in the states $X_0, \ldots, X_{N(t)-1}$. The probability distribution P_t of V_t is specified by theorem 4.2.1. The likelihood function corresponding to V_t is obtained as the Radon-Nikodym derivative of P_t with respect to a σ-finite measure σ^0, say, which does not depend on the initial distribution or the intensity rates or the parameters defining them [cf. Grenander (1981), p. 4].

If $N(t) = 0$ and $V_t = \{X_0\}$, define $S^{(0)} = S$, $\mathbb{F}^{(0)} = \mathbb{P}(S)$ the power set of S and take $\sigma^{(0)}$ to be the counting measure C on $(S^{(0)}, \mathbb{F}^{(0)})$. If $N(t) = n > 0$, introduce

$$S^{(n)} = \prod^{n} (S \times \mathbb{R}^{+}) \times S ,$$

where $\prod^{n} (S \times \mathbb{R}^{+})$ represents the n-fold Cartesian product of $S \times \mathbb{R}^{+}$ with itself. Let $\mathbb{B}^{+}$ denote the σ-field of Borel subsets of $\mathbb{R}^{+}$, ℓ denote the Lebesgue measure on $\mathbb{B}^{+}$ and define $\mathbb{F}^{(n)}$ to be the σ-field generated by subsets

$$B = \{(i_1, (a_1, b_1]), \ldots, (i_n, (a_n, b_n]), i_{n+1}\}$$

of $S^{(n)}$, where $i_1, \ldots, i_{n+1} \in S$, $(a_j, b_j] \subset \mathbb{R}^{+}$. Let

$$\sigma^{(n)}(B) = \prod_{j=1}^{n} C\{i_j\} (b_j - a_j),$$

and continue to denote by $\sigma^{(n)}$ the measure on $\mathbb{F}^{(n)}$ which is the extension of $\sigma^{(n)}$ to $\mathbb{F}^{(n)}$. One should now verify that the finiteness of the counting measure on $\mathbb{P}(S)$ and the σ-finiteness of the Lebesgue measure ℓ on $(\mathbb{R}^{+}, \mathbb{B}^{+})$, imply that $\sigma^{(n)}$ is also σ-finite on $\mathbb{F}^{(n)}$ [cf. Royden (1968), p. 265].

One can use theorem 4.2.1 to compute, for $n > 0$

$\Pr[N(t) = n]$

$$= \sum \Pr[[N(t) = n] \cap \bigcap_{j=0}^{n} [X_j = x_j] \cap \bigcap_{j=0}^{n-1} [T_j \leq t] \cap [\sum_{j=0}^{n-1} T_j \leq t]],$$

where the sum extends over all $x_j \in S$, $j = 0, 1, \ldots, n$.

Observe that, for $N(t) = n > 0$, $S^{(n)}$ is the range space of the random vector

$$\{(X_0, T_0), \ldots, (X_{n-1}, T_{n-1}), X_n\}$$

and that any set $H \in \mathbb{F}^{(n)}$ can be expressed as a finite or countable sum of the sets of the type

$$B = (\{x_0\} \times B_0, \ldots, \{x_{n-1}\} \times B_{n-1}, \{x_n\})$$

where $x_0, \ldots, x_n \in S$, and $B_0, \ldots, B_{n-1} \in \mathbb{B}^+$. Let $B(x_0, \ldots, x_n)$ denote the section of B at $(x_0, \ldots, x_n)$. It is easy to see that

$$\sigma^{(n)}(B) = \ell^{(n)}\{B(x_0, \ldots, x_n)\} \sum_{j=0}^{n} C\{x_j\}$$

where $\ell^{(n)}$ denotes the $(n+1)$ dimensional Lebesgue measure. Thus the conditional probability $P_t^{(n)}(B)$ of B given $N(t) = n$, is

$$P_t^{(n)}(B) = \Pr[B \mid N(t) = n]$$

$$= \{\Pr[N(t) = n]\}^{-1}\, p(x_0)\, \exp\{-q(x_n)t\}$$

$$\times \int_{S^{(n)} \cap B(x_0,\ldots,x_n)} \prod_{j=0}^{n-1} q(x_j, x_{j+1}) \exp[-\{q(x_j) - q(x_n)\}t_j]\, d\bar{t}_j,$$

where the integral on the right is the appropriate Lebesgue integral and the notation introduced in sections 4.1 and 4.2 is used. It follows immediately that $P_t^{(n)}$ is absolutely continuous with respect to $\sigma^{(n)}$ in the sense that whenever $\sigma^{(n)}(B) = 0$, $P_t^{(n)}(B) = 0$.

Define $S^0 = \bigcup_{n=0}^{\infty} S^{(n)}$ and

$$\mathbb{F}^0 = \{B \mid B \subset S^0,\ B \cap S^{(n)} \in \mathbb{F}^{(n)},\ n \in Z^+\}.$$

Let $B \in \mathbb{F}^0$, and let

$$\sigma^0(B) = \sum_{n=0}^{\infty} \sigma^{(n)}(B \cap S^{(n)}).$$

It is not difficult to verify that $\mathbb{F}^0$ is a σ-field of subsets of S^0 and that σ^0 is a σ-finite measure on $(S^0, \mathbb{F}^0)$. Finally, let $\mathbf{S}$ denote set of all realizations

$$\{X(u,\omega),\ 0 \le u \le t,\ \omega \in \Omega\}$$

almost all [P-measure] of which are elements of S^0. Let

$$\mathbf{IF} = \{B \mid B \subset \mathbf{S},\ B \cap S^0 \in \mathbb{F}^0\}$$

and for $B \in \mathbf{IF}$, define

$$\boldsymbol{\sigma}(B) = \sigma^0(B \cap S^0) .$$

Once again one should verify that **IF** is a σ-field of subsets of **S** and that $\boldsymbol{\sigma}$ is a σ-finite measure on $(\mathbf{S}, \mathbf{IF})$. Our preparation for the statement and proof of the following theorem is now complete.

<u>Theorem 1</u> : Let $\{X(t), t \in IR^+\}$ be a finite Markov process on the complete probability space (Ω, IF, P) and let $(\mathbf{S}, \mathbf{IF}, \boldsymbol{\sigma})$ be the measure space defined above. Then for any $B \in \mathbf{IF}$,

$$\Pr[V_t \in B] = \int_B L(v, Q)\, d\boldsymbol{\sigma}(v),$$

where

$$L(v, Q) = \begin{cases} p(x_0) \exp\{-q(x_0)t\} , & \text{if } v = \{x_0\} , \\ p(x_0) \exp\{-q(x_n)t\} \\ \quad \times \prod_{j=0}^{n-1} q(x_j, x_{j+1}) \exp[-\{q(x_j) - q(x_n)\}t] , & (1) \\ \quad \text{if } v = \{(x_0, t_0), \ldots, (x_{n-1}, t_{n-1}), x_n\} \\ \quad \text{with } n>0,\ t_j \geq 0,\ j = 0,1,\ldots,(n-1),\ \sum_{j=0}^{n-1} t_j \leq t , \\ 0 , & \text{otherwise} . \end{cases}$$

<u>Proof</u> : We have already seen above that the conditional distribution $P_t^{(n)}$ of V_t, given $N(t) = n$, is absolutely continuous with respect to $\sigma^{(n)}$. Thus by the Radon-Nikodym theorem, there exists

an almost everywhere $(\sigma^{(n)})$ - unique and non-negative function $g_n(x_0, \ldots, x_n, t_0, \ldots, t_{n-1})$ on $S^{(n)}$ such that for all $B \in \mathbb{F}^{(n)}$

$$P_t^{(n)}(B) = \int_B g_n \, d\sigma^{(n)} .$$

Let $x_0, \ldots, x_n \in S$ be fixed and let

$$F_n(x_0, \ldots, x_n, t_0, \ldots, t_{n-1})$$

$$= P_t^{(n)}[X_0 = x_0, \ldots, X_n = x_n, T_0 \leq t_0, \ldots, T_{n-1} \leq t_{n-1}].$$

If F has partial derivative

$$\partial^n F_n(x_0, \ldots, x_n, t_0, \ldots, t_{n-1})/\partial t_0 \, \partial t_1, \ldots, \partial t_{n-1}$$

of order n, it equals g_n almost everywhere $(\sigma^{(n)})$. It follows from theorem 4.2.1 that

$$g_n = L(v, Q)/Pr[N(t) = n]$$

with $L(., Q)$ defined by (1). Hence for any $B \in \mathbf{IF}$,

$$Pr[V_t \in B] = \sum_{n=0}^{\infty} Pr[N(t) = n] \int_{B \cap S^{(n)}} g_n \, d\sigma^{(n)}$$

$$= \sum_{n=0}^{\infty} \int_{B \cap S(n)} L(., Q) \, d\sigma^{(n)}$$

$$= \int_{B \cap \mathbf{S}} L(., Q) \, d\boldsymbol{\sigma}$$

$$= \int_B L(., Q) \, d\boldsymbol{\sigma} .$$

This completes the proof of the theorem, [cf. Albert (1962)].

Remark : Let P_t denote the probability measure induced on $(\mathbf{S}, \mathbb{F})$ by $\{X(u), 0 \leq u \leq t\}$. The above theorem asserts that P_t is absolutely continuous with respect to σ and that $L(., Q)$ is a Radon-Nikodym derivative of P_t with respect to σ. Observe that $L(.,.)$ is a function on $\mathbf{S} \times \Theta$, where Θ is the set

$$\{q(i,j) \mid i, j \in S, \; q(i,j) \geq 0, \; i \neq j, \; q(i,i) = \sum_{j \neq i} q(i,j)\}$$

of the intensity rates. If we fix $v \in \mathbf{S}$, then $L(v, .)$ as a function on Θ, is the likelihood function or simply the likelihood of Q corresponding to the observation v. In order to avoid trivial complications of notation, in future we shall replace v by V_t in $L(v, Q)$ without causing any confusion.

3. ML ESTIMATION

Suppose that a finite Markov process with state-space $S = \{1, \ldots, M\}$, known initial distribution $\{p(i), i \in S\}$ and unknown intensity rates $q(i, j), i \neq j, i, j \in S$ is observed completely over $[0, t]$. Then the likelihood $L(V_t, Q)$ of Q corresponding to observation $V_t = \{X(u), 0 \leq u \leq t\}$ is expressible in the following form :

$$L(V_t, Q) = \prod_{i=1}^{M} \{p(i)\}^{\delta(X(0),i)} \prod_{i=1}^{M} \prod_{j \neq i} \{q(i,j)\}^{N_t(i,j)}$$

$$\times \exp\{- \sum_{i=1}^{M} A_t(i)\, q(i)\}, \qquad (1)$$

where $q(i,j)$, $q(i) = \sum_{j \neq i} q(i,j)$, $i \neq j$, $i, j \in S$ are the unknown parameters, $\delta(i,j)$ is the Kronecker delta, $N_t(i, j)$ is the number of direct transitions from i to j, and $A_t(i)$ is the total time spent by the process in state i during the period of observation. The ML estimators of $q(i,j)$ are statistics $\hat{q}(i,j,t)$ with values in the parameter space Θ, such that

$$L(V_t , \hat{Q}(t)) = \sup \{L(V_t, Q) \mid q(i, j) \in \Theta , i, j \in S\} ,$$

where $\hat{Q}(t) = ((\hat{q}(i,j,t)))$ is the matrix Q with $q(i,j)$ replaced by $\hat{q}(i,j,t)$, $i, j \in S$.

In order to obtain $\hat{q}(i,j,t)$ from the expression (1) for the likelihood, we first observe that $L(V_t, Q)$ and $\ell(V_t , Q) = \log L(V_t, Q)$ being monotone functions of each other, it is enough to search for $\hat{Q}(t)$ which maximizes $\ell(V_t, Q)$. A general method for obtaining $Q(t)$, when $\ell(V_t, Q)$ is twice differentiable with respect to $q(i, j)$, is to obtain a solution of the likelihood equations.

$$\partial\ell(V_t, Q)/\partial q(i,j) = 0 , \quad i \neq j, i, j \in S \tag{2}$$

and then verify that at the solution the Hessian matrix $((\partial^2 \ell(V_t, Q)/\partial q(i, j) \partial q(r, s)))$ is a.s. negative definite. We adopt this method to obtain the ML estimators of $q(i, j)$ in the following

<u>Lemma 1</u> : Suppose the finite Markov process $\{X(t), t \in \mathbb{R}^+ \}$ is irreducible at the true parametric point,

$$\{q^0(i, j) \quad i \neq j, \quad i, j \in S\} \quad \varepsilon \quad \Theta .$$

Suppose, further that the set

$$D = \{(i, j) \mid i \neq j \ , \quad q^0(i, j) > 0, \quad i, j \in S\}$$

is known.

Then for all sufficiently large t, the ML equations (2) admit the unique solution

$$\hat{q}(i,j,t) \quad = \quad N_t(i,j)/A_t(i), \quad (i, j) \in D \tag{3}$$

with P_0-probability one, where P_0 corresponds to the probability measure corresponding to the true parametric point $\{q^0(i,j),(i,j) \in D\}$ Furthermore, with P_0-probability one, the solution (3) provides a maximum of the likelihood for all sufficiently large t.

Proof : Observe that a formal differentiation of

$$\ell(V_t, Q) = \delta(X(0), i) \log p(i) + \sum_{(i,j) \in D} N_t(i,j) \log q(i,j)$$

$$- \sum_{(i,j) \in D} A_t(i) \, q(i,j)$$

with respect $q(i,j)$, $(i,j) \in D$, yields the equations

$$N_t(i,j) \quad - \quad A_t(i) \, q(i,j) \quad = 0, \ (i,j) \in D.$$

By the assumption of irreducibility of the Markov process at $q^0(i,j)$ $(i,j) \in D$, both $N_t(i,j)$ and $A_t(i)$ are positive with P_0-probability one for all sufficiently large t and therefore (3) is the unique solution of (2).

Observe further that for $(i, j), (r, s) \varepsilon D$,

$$[\partial^2 \ell(V_t, Q)/\partial q(i,j)\, \partial q(r,s)]_{\hat{q}(i,j,t),\ \hat{q}(r,s,t)}$$

$$= - \{A_t^2(i)/N(i,j)\}\ \delta(i,j;\ r,s) ,$$

where $\delta(i,j;\ r,s)$ is the bivariate Kronecker delta defined by (4.2.15). If d denotes the number of elements in the set D, then the almost sure positive nature of $N_t(i,j)$ and $A_t(i)$ for all sufficiently large t, implies that the dxd matrix $((\partial^2(V_t,Q)/\partial q(i,j)\ \partial q(r,s)))$ evaluated at $\hat{q}(i,j,t),\ i, j \varepsilon D$ is almost surely negative definite for all sufficiently large t and therefore the estimators (3) are ML estimators with probability one for all sufficiently large t.

Corollary 1 : The ML estimator of $q(i)$ is

$$\hat{q}(i, t) = \sum_{j\varepsilon D_i} \hat{q}(i,j,t) ,\ i \varepsilon S ,$$

where

$$D_i = \{(j) \mid j \varepsilon S, (i,j) \varepsilon D\} .$$

Proof : This is a striaghtforward consequence of the well-known principle of invariance for ML estimation [cf. Zacks (1971),p.223].

In the absence of any precise knowledge about the joint distribution of $N_t(i,j)$ and $A_t(i)$, $(i,j) \varepsilon D$, it is almost impossible to study the distributional properties of the ML estimators of $q(i,j)$ for finite t. We,therefore,proceed to study in the next section,their asymptotic properties as $t \to \infty$.

4. ASYMPTOTIC PROPERTIES OF ML ESTIMATORS

Let $\{X(t),\ t \in \mathbb{R}^+\}$ be an irreducible Markov process with state-space $S = \{1, \ldots, M\}$. Assume that the true intensity rates $q^0(i,j)$ are positive for all $i, j \quad S, i \neq j$. This is equivalent to the assumption that the set D introduced in lemma 3.1 is

$$D = S \times S - \{(i, i),\ i \in S\}. \tag{1}$$

When D is a proper subset of the one on the right of (1). only minor modifications are required in the arguments that follow. We shall also suppress the superscript 0 in $q^0(i,j)$ and write $q(i,j)$ for $q^0(i, j)$. The methods adopted to establish the consistency and asymptotic normality of ML estimators are minor modifications of those of Grenander (1981, p. 310-317) who in turn simplifies the methods of Albert (1962).

Define ,

$$\xi_{ij}(t) = \{N_t(i,j) - A_t(i)\, q(i,j)\}/\sqrt{t}, \quad i \neq j, \quad i, j \in S$$

$$\xi_{jj}(t) \equiv 0, \quad j \in S \tag{2}$$

and let $\mathbf{W}$ be a set of auxiliary variates w_{ij} with $w_{jj}=0, i,j \in S$. The following theorem obtains the joint moment generating function (m.g.f.).

$$\varphi(\mathbf{W}, t) = E[\exp\{-\sum_{i=1}^{M}\sum_{j=1}^{M} w_{ij}\, \xi_{ij}(t)\}]$$

of the r.v.s $\xi_{ij}(t),\ i, j \in S$. Let E_{rs} denote as usual, the rxs matrix whose every element is equal to unity. Albert (1962) has introduced the following analogue of inner product of two vectors. If A and B are two matrices of the same order

$$< A, B> = \sum_i \sum_j a_{ij} b_{ij} .$$

In this notation

$$\varphi(\mathbf{W}, t) = E[\exp\{ -< \mathbf{W}, \boldsymbol{\xi}(t)> \}]$$

where $\boldsymbol{\xi}(t) = ((\xi_{ij}(t)))$, $i, j = 1, \ldots, M$.

<u>Theorem 1</u> : Let $\{ X(t), t \in \mathbb{R}^+\}$ be an irreducible finite Markov process with every $q(i,j) > 0$, $i \neq j$, $i, j \in S$. Define the M-square matrix $\mathbf{R} = ((r_{ij}))$ as

$$r_{ii} = - q(i) + \sum_{u \neq i} q(i, u) w_{ij}/\sqrt{t} , \quad i \in S ;$$

$$r_{ij} = q(i,j) \exp \{- w_{ij}/\sqrt{t} \} , \quad i \neq j, \; i, j \in S. \tag{3}$$

Let $p(0) = (p_1(0), \ldots, p_M(0))^T$ be the initial probability vector. The joint m.g.f. $\varphi(\mathbf{W}, t)$ is

$$\varphi(\mathbf{W}, t) = \langle \exp (t\mathbf{R}^T) p(0), E_{M1} \rangle . \tag{4}$$

<u>Proof</u> : Define

$$\Phi(t) = \exp[- \Sigma \Sigma \nu_{uv} \{N_t(u,v) - q(u,v) A_t(u)\}$$

where ν_{uv}, $u, v \in S$, $\nu_{uu} = 0$, $u \in S$, are new auxiliary variates. Let

$$\varphi_k(t) = E[\Phi(t)\, I[X(t) = k]]\ , \quad k \in S\ .$$

Let $h > 0$ and observe that

$$\varphi_k(t+h) = E[E\{\Phi(t+h)\, I[X(t+h) = k] \mid \mathbb{F}_t\}] \qquad (5)$$

where $\mathbb{F}_t$ denotes the σ-field $\sigma\,\{X(u),\ 0 \le u \le t\}$ induced by the r.v.s $X(u),\ 0 \le u \le t$. It is easy to see that

$$\Phi(t+h) = \Phi(t) \exp\,[-\Sigma\ \nu_{uv}\{N_{t+h}(u,v) - N_t(u,v)$$

$$- q(u,v)\{A_{t+h}(u) - A_t(u)\}\}]$$

$$= \Phi(t)\, Z(t, t+h)\ , \quad \text{say}\ .$$

Since $\Phi(t)$ is $\mathbb{F}_t$ - measurable, we can rewrite (5) as

$$\varphi_k(t+h) = E[\Phi(t)\, E\{Z(t, t+h)\, I[X(t+h) = k] \mid \mathbb{F}_t\}]$$

$$= E[\Phi(t)\, E\{Z(t, t+h)\, I[X(t+h) = k] \mid X(t)\}]\ ,$$

which is a consequence of the Markov property. Since $X(t)$ is a discrete r.v., we have the further simplification :

$$\varphi_k(t+h) = \sum_{j\in S} E[\Phi(t)\, I[X(t) = j] E\{Z(t,t+h)\, I[X(t+h) = k] \mid X(t) = j\}].$$

Thus, it is necessary to obtain

$$E\{Z(t, t+h)\, I[X(t+h) = k] \mid X(t) = j\}\ ,$$

in evaluating which we need to consider the values of $Z(t, t+h)$ on the set $[X(t) = j,\ X(t+h) = k]$ only.

If $j = k$, there is no jump in $(t, t+h]$ with probability

$1 - q(k)h + o(h)$ and then

$$N_{t+h}(u,v) - N_t(u,v) \equiv 0,\ A_{t+h}(u) - A_t(u) = h\ \delta_{jk},\ u,\ v \in S.$$

Thus when $j = k$, $X(t) = X(t+h) = k$, and there is no jump in $(t, t+h]$,

$$Z(t, t+h)\ I[X(t+h) = k] = \exp\{h \sum_{v \neq k} \nu_{kv}\ q(k,v)\}\ .$$

If $j \neq k$, and there is only one change of state during $(t, t+h]$, which happens with probability $q(j, k)h + o(h)$,

$$N_{t+h}(j, k) - N_t(j, k) = 1,$$

$$N_{t+h}(u, v) - N_t(u,v) = 0 \quad \text{for all other } u,\ v \in S,$$

and $$A_{t+h}(u) - A_t(u) = o(1) \quad \text{for all} \quad u \in S\ .$$

Thus on the set $[X(t) = j\ ,\ X(t+h) = k]\ ,\ j \neq k$,

$$Z(t, t+h)\ I[X(t+h) = k] = \exp\{- \nu_{jk} + o(1)\}$$

with probability $q(j,k)h + o(h)$. Of course, we can have $X(t) = j$ and $X(t+h) = k$ with multiple changes of state during $(t, t+h]$; a situation which has probability $o(h)$, $h \downarrow 0$.

Hence, if $j = k$

$$E[Z(t, t+h)\ I[X(t+h) = k] \mid X(t) = k]$$

$$= \exp\{ h \sum_{v \neq k} \nu_{kv}\ q(k,v)\}\ \{1 - q(k)h + o(h)\} + o(h)\ ;$$

and if $j \neq k$,

$$E[Z(t,\ t+h)\ I[X(t+h) = k] \mid X(t) = j]$$

$$= \exp\{-\nu_{jk} + o(1)\}\ q(j,\ k)h + o(h)\ .$$

Thus

$$\varphi_k(t+h) = \varphi_k(t)\ \exp\{h \sum_{v \neq k} \nu_{kv}\ q(k,v)\}\ \{1 - q(k)h + o(h)\}$$

$$+ \sum_{u \neq k} \varphi_k(t)\ \exp\{-\nu_{uk} + o(1)\}\ q(u,k)h + o(h).$$

Alternatively, the above equation can be rewritten as

$$\varphi_k(t+h) = \varphi_k(t)\ \{1 + h \sum_{v \neq k} \nu_{kv}\ q(k,v) - q(k)h\}$$

$$+ \sum_{u \neq k} \varphi_u(t)\ \exp(-\nu_{uk})\ q(u,k)h + o(h)\ .$$

Following the usual technique [cf. theorem 2.2.2] , one has

$$d\varphi_k(t)/dt = \{-q(k) + \sum_{v \neq k} \nu_{kv}\ q(k,v)\}\ \varphi_k(t)$$

$$+ \sum_{u \neq k} q(u,k)\ \exp(-\nu_{uk})\ \varphi_u(t),\quad k \in S. \tag{6}$$

Let now

$$\ell_{uv} = q(u,v)\ \exp\{-\nu_{uv}\}\ ,\quad u \neq v\ ,$$

$$\ell_{uu} = -q(u) + \sum_{v \neq u} \nu_{uv}\ q(u,v),\quad u \in S$$

and define Λ to be the M-square matrix $((\ell_{uv}))$. If $\Psi(t)$ denotes the column vector $(\varphi_1(t),\ \ldots,\ \varphi_M(t))$, the M equations (6) can be combined into the single matrix equation

$$d\Psi(t)/dt = \Lambda^T\Psi(t) \tag{7}$$

with the obvious initial condition $\Psi(0) = p(0)$. The matrix equation (6) has the unique solution

$$\Psi(t) = \exp\{ t\Lambda^T\} p(0)$$

and since $\varphi(\nu, t) = \sum_{u=1}^{M} \varphi_u(t)$, it follows that the joint m.g.f.of

$$N_t(u,v) - q(u,v) A_t(u) , \quad u, v \in S \tag{8}$$

is given by

$$< \exp (t\Lambda^T) p(0) , E_{M1} > . \tag{9}$$

Now observe that the joint m.g.f. of $\xi_{ij}(t)$, i, j ε S, is given by $\varphi(\nu/\sqrt{t} , t)$. The proof of the theorem is completed by replacing ν_{uv} by $w_{uv}/\sqrt{t}$, u, v ε S in the expression (9) for the joing m.g.f. of the r.v.s specified by (8).

The above theorem can be used to obtain the joint asymptotic distribution of the r.v.s $\xi_{ij}(t)$ defined by (1). This we proceed to do under the assumption that the intensity matrix Q has simple characteristic roots $\lambda_1 = 0$, $\lambda_2, \ldots, \lambda_M$.

Let η_j and ξ_j denote the left and right characteristic vectors of Q, corresponding to the roots λ_j, j = 1,..., M, such that

$$\eta_1 = (\pi_1, \ldots, \pi_M)^T , \quad \xi_1 = E_{M1}$$
$$\eta_j^T \xi_k = \delta_{jk} , \quad j, k = 1,\ldots, M, \tag{10}$$

where $\pi = (\pi_1,\ldots, \pi_M)^T$ is the invariant probability vector of the process. Let V and U denote the M-square matrices $(\eta_1, \ldots, \eta_M)$ and $(\xi_1, \ldots, \xi_M)$ respectively. Then $V^T U = I$ and the spectral resolution of Q is

$$Q = U \Lambda V^T , \tag{11}$$

where Λ is the diagonal matrix diag $(\lambda_1, \ldots, \lambda_M)$.

Theorem 2 : If the intensity matrix Q has simple characteristic roots, then

$$\lim_{t\to\infty} \varphi(\mathbf{W}, t) = \exp \{ \frac{1}{2} \sum_{i=1}^{M} \sum_{j=1}^{M} \pi_i \, q(i,j) \, w_{ij}^2 \} \tag{12}$$

and therefore, the r.v.s $\xi_{ij}(t)$ are asymptotically, as $t \to \infty$, distributed like independent normal $N(0 , \pi_i \, q(i,j))$ variates, $i \neq j$, $i, j \in S$.

Proof : Expanding the expression (3) for r_{ij} , $i \neq j$, one has

$$r_{ij}(t) = q(i,j) \{1 - w_{ij}/\sqrt{t} + w_{ij}^2/(2t) + o(1/t)\}, \tag{13}$$

where we write $r_{ij}(t)$ for r_{ij} to emphasise its dependence on t.

Define

$$a_{ij} = \begin{cases} \sum_{v \neq i} q(i, v) \, w_{iv} & \text{if } i = j , \\ - q(i, j) \, w_{ij} & \text{if } i \neq j, \; j \in S, \end{cases}$$

and $b_{ij} = \frac{1}{2} q(i,j) \, w_{ij}^2 \, \{1 - \delta_{ij}\}$, $i, j \in S$.

Introducing the M-square matrices $A = ((a_{ij}))$ and $B = ((b_{ij}))$ we can rewrite the matrix $R(t)$ as

$$R(t) = Q + A/\sqrt{t} + B/t + o(1/t) , \tag{14}$$

where, and in what follows, $o(1/t)$ represents a matrix of appropriate dimensions and whose every element is $o(1/t)$.

Since the matrices U and V introduced just after equation (10), are non-singular and $V^T U = I_M$, we have

$$\begin{aligned}\{V^{-1} R^T(t) (U^T)^{-1}\}^2 &= V^{-1} R^T(t) (U^T)^{-1} V^{-1} R^T(t)(U^T)^{-1} \\ &= V^{-1} R^T(t) V V^{-1} R^T(t) (U^T)^{-1} \\ &= V^{-1} \{R^T(t)\}^2 (U^T)^{-1} ,\end{aligned}$$

so that for all $n \geq 1$

$$\{V^{-1} R^T(t) (U^T)^{-1}\}^n = V^{-1}\{R^T(t)\}^n (U^T)^{-1} .$$

Thus

$$\exp\{t V^{-1} R^T(t) (U^T)^{-1}\} = V^{-1} \exp\{t R^T(t)\} (U^T)^{-1}$$

or equivalently

$$\exp \{t R^T(t)\} = V \exp \{t V^{-1} R^T(t) (U^T)^{-1}\} U^T .$$

Thus the joint m.g.f. $\varphi(\mathbf{W} , t)$ of $\xi_{ij}(t)$ $i \neq j$, $i, j \in S$ is expressible as

$$\begin{aligned}\varphi(\mathbf{W} , t) &= E_{1M} \exp\{ t R^T(t)\} p(0) \\ &= E_{1M} V \exp \{S^T(t)\} U^T p(0) \\ &= \alpha^T \exp \{S^T(t)\} \beta ,\end{aligned} \tag{15}$$

where $\alpha^T = E_{1M}V$, $\beta = U^T p(0)$ and

$$S(t) = t\, V^{-1}\{R^T(t)\,(U^T)^{-1}\}^T = t\, U^{-1}\, R(t)\,(V^T)^{-1} .$$

Substituting for $R(t)$ from (14) in the above expression, one has

$$S(t) = t\, U^{-1}\, Q(V^T)^{-1} + \sqrt{t}\, U^{-1}A(V^T)^{-1} + U^{-1}B(V^T)^{-1} + o(1)$$

$$= t\,\Lambda + \sqrt{t}\; V^T A\, U + V^T B\, U + o(1) ,$$

a simplification obtained by using (11) and the fact that $V^T U = I_M$.

Let $s_{ij}(t)$ denote the (i,j) element of the matrix $S(t)$. Since $\lambda_1 = 0$ and $A\xi_1 = 0$,

$$s_{11}(t) = (V^T B\, U)_{11} + o(1)$$

where $(V^T B\, U)_{11}$ represents the (1,1) element of $V^T B\, U$. Similarly,

$$s_{jj}(t) = t\lambda_j + o(t) , \quad j = 2,\ldots, M$$

and when $i \neq j$, $s_{ij}(t) = o(t)$. Combining these fact, we have

$$S(t) = s_{11}(t)\, I_M + t\, H(t),$$

where $H(t)$ is an M-square matrix whose diagonal elements are the diagonal elements of $\Lambda + o(1)$ and whose off-diagonal elements are $o(1)$. Hence

$$\lim_{t\to\infty} H(t) = \Lambda .$$

One consequence of this result is that the characteristic roots of $H(t)$ converge, as $t \to \infty$, to the characteristic roots of Λ, which are $\lambda_1 = 0, \lambda_2, \ldots, \lambda_M$ and are distinct by assumption. Thus for sufficiently large t, the characteristic roots of $H(t)$ are also distinct. One, therefore, has the spectral representation, valid for all sufficiently large t for $H(t)$:

$$H(t) = U(t) \; \Delta(t) \; V^T(t)$$

where $U(t)$ and $V(t)$ are the matrices formed by the right and left characteristic vectors of $H(t)$ and $\Delta(t)$ is the diagonal matrix of characteristic roots of $H(t)$. Moreover, it is possible to show that as $t \to \infty$, $\Delta(t) \to \Lambda$, $U(t) \to U_0$, $V(t) \to V_0$, where U_0 and V_0 are the matrices formed by the right and left characteristic vectors of Λ, [cf. Kato (1966), p. 107] . Since Λ is a diagonal matrix of distinct elements, $U_0 = V_0 = I_M$. It follows that

$$\begin{aligned} \exp\{t\,H(t)\} &= U(t) \; \exp\{t\,\Delta(t)\} \; V^T(t) \\ &\to U_0 \; \mathrm{diag}\,(1, 0, \ldots, 0) \; V_0 \end{aligned}$$

since $\mathrm{Re}(\lambda_j) < 0$ for $j = 2,3,\ldots, M$ [cf. lemma 2.4.2].

In particular,

$$\lim_{t\to\infty} \; \exp\{t\,S^T(t)\} = \exp\{c\,I_M\} \; \mathrm{diag}\,[1, 0, \ldots, 0]$$

where $c = \lim s_{11}(t) = (V^T\,B\,U)_{11}$. But observe that

$$\begin{aligned} (V^T B\,U)_{11} &= \eta_1^T \, B \, \xi_1 = (\pi_1, \ldots, \pi_M) \, B \, E_{M1} \\ &= \frac{1}{2} \sum_{i=1}^{M} \sum_{j\neq i} \pi_i \, q(i,j) \, w_{ij}^2 \, . \end{aligned}$$

The required result (12) follows from (15) on noting that first elements of both α and β are unity.

In order to obtain the joint asymptotic distribution of the ML estimators $\hat{q}(i,j,t)$ of $q(i,j)$, we need the following

Lemma 1 : If the intensity matrix Q has simple characteristic roots, then as $t \to \infty$,

$$\begin{aligned} N_t(i,j)/t &\xrightarrow{P} \pi_i q(i,j), \quad i \neq j, \\ A_t(i)/t &\xrightarrow{P} \pi_i \quad , \quad i, j \in S, \end{aligned} \tag{16}$$

where $\xrightarrow{P}$ denotes convergence in probability.

Proof : This is an easy consequence of corollary 4.2.2 and Chebychev's inequality.

Corollary 1 : If the intensity matrix Q has simple characteristic roots, then the estimators $\hat{q}(i,j,t)$ of $q(i,j)$ are weakly consistent and the joint distribution of

$$\sqrt{t}\,\{\hat{q}(i,j,t) - q(i,j)\}, \quad i \neq j, \ i, \ j \in S$$

converges to the joint distribution of independent r.v.s Z_{ij} with normal $N(0, q(i,j)/\pi_i)$ distribution.

Proof : Observe that

$$\sqrt{t}\{\hat{q}(i,j,t) - q(i,j)\} = \xi_{ij}(t)/\{A_t(i)/t\}, \quad i \neq j, \ i,j \in S,$$

where $\xi_{ij}(t)$ are defined by (2). The corollary follows by theorem 2, lemma 1 and an application of Slutsky's theorems [cf. Rao (1973), p. 122] .

The results of this section can be described as those pertaining to the non-parametric estimation related to finite Markov processes, in view of the fact that a specification of all the intensity rates is sufficient for obtaining its family of finite dimensional distributions. In the next section we discuss the results of Ranneby (1978) about the strong consistency of the approximate ML estimators of the parameters defining the intensity rates.

5. STRONG CONSISTENCY OF THE ML ESTIMATORS

Suppose that the intensity rates of a finite irreducible Markov process depend on a vector-valued parameter $\boldsymbol{\theta} = (\theta_1, \ldots, \theta_k) \varepsilon \Theta$, the parameter space which is assumed to be a compact subset of the k-dimensional Euclidean space. If $V_t = \{X(u), 0 \leq u \leq t\}$ is an observation on the Markov process over the time interval $[0, t]$, then the ML estimator $\hat{\boldsymbol{\theta}}(t)$ of $\boldsymbol{\theta}$ based on V_t is an k-dimensional statistic with values in Θ such that

$$L(V_t, \hat{\boldsymbol{\theta}}(t)) = \sup\{L(V_t, \boldsymbol{\theta}) \mid \boldsymbol{\theta} \varepsilon \Theta\}, \qquad (1)$$

where $L(V_t, \boldsymbol{\theta})$ is specified by equation (2.1) with $q(i,j)$ replaced by $q(i,j,\boldsymbol{\theta})$ to make the dependence of $q(i,j)$ on $\boldsymbol{\theta}$ explicit. In general, an ML estimator satisfying (1) may not exist. We therefore define a c-approximate ML estimator $\hat{\boldsymbol{\theta}}_c(t)$ as a k-dimensional statistic with values in Θ such that

$$L(V_t, \hat{\boldsymbol{\theta}}_c(t)) \geq c \sup\{L(V_t, \boldsymbol{\theta}) \mid \boldsymbol{\theta} \varepsilon \Theta\} \qquad (2)$$

for an arbitrary but fixed c, $0 < c < 1$. Such an approximate ML estimator always exists for every fixed $c \varepsilon (0,1)$, although it may not be unique, e.g., if $c' > c$, a c'-approximate ML estimator is also a c-approximate ML estimator. In this section we follow

Ranneby (1978) to establish that c-approximate ML estimator is strongly consistent in the sense that as $t \to \infty$, $\hat{\theta}_c(t) \to \theta^0$ with P_{θ^0} - probability one. Here θ^0 denotes the true parametric point and P_{θ^0} is the probability measure governing the finite Markov process when θ^0 is the true parametric point. We need the following lemmas.

<u>Lemma 1</u> : Let $\alpha_1, \ldots, \alpha_m$ and $\beta_1, \ldots, \beta_m$ be non-negative numbers such that $\sum_{i=1}^{m} \alpha_i = \sum_{i=1}^{m} \beta_i = 1$. Then

$$- 2 \sum_{i=1}^{m} \alpha_i \log (\beta_i/\alpha_i) \geq \sum_{i=1}^{m} (\alpha_i - \beta_i)^2 , \qquad (3)$$

where $\alpha_i \log (\beta_i/\alpha_i)$ is defined to be zero if $\alpha_i = \beta_i = 0$ or if $\alpha_i = 0$. If $\beta_i = 0$ and $\alpha_i > 0$, it is defined to be $-\infty$.

<u>Proof</u> : In case $\alpha_i = 0$ and/or $\beta_i = 0$, the conventions made above imply that

$$-2 \alpha_i \log(\beta_i/\alpha_i) - 2(\alpha_i - \beta_i) \geq (\alpha_i - \beta_i)^2 .$$

So let α_i and β_i be both positive. The Taylor expansion of $x \log x$ around x_0 gives

$$x \log x = x_0 \log x_0 + (x-x_0) (1 + \log x_0) + \frac{1}{2x_1} (x-x_0)^2 ,$$

where x_1 is between x and x_0 in the sense that $x_1 \in [x,x_0]$ if $x < x_0$ and $x_1 \in [x_0, x]$ if $x \geq x_0$. If x, x_0 are both at most equal to one, then $x_1 \leq 1$. One can therefore obtain

$$- 2(x \log x_0 - x \log x) - 2(x - x_0) \geq (x - x_0)^2 .$$

Replace x by α_i, x_0 by β_i and then sum over $i = 1, \ldots, M$, to obtain the required result (3) which is due to Birch (1964).

Lemma 2 : Let $a_1, \ldots, a_m$; $b_1, \ldots, b_m$ and $c_1, \ldots, c_m$ be positive real numbers. Then

$$\sum_{i=1}^{m} a_i \log (c_i/b_i) - \sum_{i=1}^{m} a_i c_i/b_i + \sum_{i=1}^{m} a_i$$

$$= - \sum_{i=1}^{m} a_i (b_i - c_i)^2/\{2z_i^2\} , \qquad (4)$$

where z_i lies between b_i and c_i, $i = 1, \ldots, m$.

Proof : Consider the Taylor expansion of $\log x$ around $x = x_0$, $x_0 > 0$, to obtain

$$\log x = \log x_0 + (x - x_0)x_0^{-1} + (x - x_0)^2 x_1^{-2}/2 , \qquad (5)$$

where x_1 lies between x and x_0. In (5), put $x = c_i/b_i$, $x_0 = 1$, $x_1 = z_i/b_i$, multiply by a_i and then sum over $i = 1, \ldots, m$ to obtain (4).

Remark : The inequality (4) is valid even if $a_i = 0$ for some i if we take the corresponding z_i to be zero.

Lemma 3 : Let $\{X(t), t \in \mathbb{R}^+\}$ be a finite, irreducible Markov process with state space $S = \{1, \ldots, M\}$. Then whatever may be the initial distribution, as $t \to \infty$,

$$A_t(i)/t \to \pi_i \quad , \tag{6}$$

$$N_t(i, j)/t \to \pi_i\, q(i, j), \tag{7}$$

$$N(t)/t = \sum_{i=1}^{M} \sum_{j\neq i} N_t(i,j)/t \to \sum_{i=1}^{M} \sum_{j\neq i} \pi_i q(i,j), \tag{8}$$

$$A_t(i)/N(t) \to \pi_i/\{ \sum_u \sum_{v\neq u} \pi_u q(u,v) \}, \tag{9}$$

$$N_t(i,j)/N(t) \to \pi_i q(i,j)/\{\sum_u \sum_{v\neq u} \pi_u\, q(u,v)\}, \tag{10}$$

$$N_t(i)/A_t(i) = \sum_{j\neq i} N_t(i,j)/A_t(i) \to q(i) \tag{11}$$

with probability one.

Proof : Suppose that the initial distribution of the process is its unique invariant distribution $\{\pi_i, i \in S\}$. Then the irreducible Markov process is a strictly stationary process which is metrically transitive [cf. Doob (1953), p. 460, 511] . Observe further that

$$A_t(i)/t = (1/t) \int_0^t I[X(u) = i]du \tag{12}$$

and that

$$N_t(i,j)/t = (1/t) \int_0^t f_{ij}(X(u-o), X(u))\, du, \tag{13}$$

where $f_{ij}(.,.)$ is a function defined on $S \times S$ by the relation

$$f_{ij}(u,v) = \begin{cases} 0 , & \text{if } u = v , \\ 1 , & \text{if } u = i,\ v = j,\ i \neq j . \end{cases}$$

It is easy to check that the integrand in (13) equals 1 only if a transition occurs from i to j at epoch u. The almost sure limits in (6) and (7) are immediate consequences of the ergodic theorem. When the initial distribution is arbitrary, some more sophisticated analysis is required for which we refer the interested reader to Azema, Kaplan-Duflo and Revuz (1967), [cf. Cinlar (1975), p. 269- 270 also] .

The rest of the results (8) - (11) are simple consequences of (6) and (7).

We now state a modification of the identifiability condition introduced by Rao (1973) in connection with ML estimation for multinomial populations.

<u>Identifiability condition</u> : Let $\{ \boldsymbol{\theta}_n, \ n \geq 1\}$ be a sequence of parametric points in Θ such that

$$\lim_{n\to\infty} q(i,j, \boldsymbol{\theta}_n) = q(i,j, \boldsymbol{\theta}^0) \tag{14}$$

for all $i \neq j$, $i, j \in S$ and some fixed $\boldsymbol{\theta}^0$ in Θ . Then

$$\lim_{n\to\infty} \boldsymbol{\theta}_n = \boldsymbol{\theta}^0 . \tag{15}$$

We are now in a position to establish the following

Theorem 1 : Suppose for each $\boldsymbol{\theta} \in \Theta$, the finite Markov process is irreducible and suppose that the identifiability condition holds. Then any c-approximate ML estimator is strongly consistent for $\boldsymbol{\theta}$ as $t \to \infty$.

Proof : Suppose that the initial distribution of the Markov process also depends on $\boldsymbol{\theta}$ and let $p_i(\boldsymbol{\theta}) = \Pr[X(0) = i]$, $i \in S$. A slight re-arrangement of the terms in equation (2.1), yields the following expression for the log-likelihood $\ell_t(\boldsymbol{\theta}) = \log L(V_t, \boldsymbol{\theta})$:

$$\ell_t(\boldsymbol{\theta}) = \log p_{X(0)}(\boldsymbol{\theta}) + \sum_{i=1}^{M} \sum_{j\neq i} N_t(i,j) \log r_{ij}(\boldsymbol{\theta})$$

$$+ \sum_{i=1}^{M} \{N_t(i) \log q(i, \boldsymbol{\theta}) - A_t(i)\, q(i, \boldsymbol{\theta})\}, \quad (16)$$

where $r_{ij}(\boldsymbol{\theta}) = q(i, j, \boldsymbol{\theta})/q(i,\boldsymbol{\theta})$ is the one-step transition probability associated with the jump Markov chain $\mathbf{X}(\mathbb{R})$.

Observe that

$$\sum_{i=1}^{M} \sum_{j\neq i} N_t(i,j) \log r_{ij}(\boldsymbol{\theta})$$

$$= \sum_{i=1}^{M} N_t(i) \sum_{j\neq i} \{N_t(i,j)/N_t(i)\} \log[r_{ij}(\boldsymbol{\theta})/\{N_t(i,j)/N_t(i)\}]$$

$$+ \sum_{i=1}^{M} N_t(i) \sum_{j\neq i} \{N_t(i,j)/N_t(i)\}\log\{N_t(i,j)/N_t(i)\}, \quad (17)$$

where we interpret $N_t(i,j)/N_t(i)$ as zero if $N_t(i) = 0$. By virtue of lemma 1,

$$\sum_{j\neq i} \{N_t(i,j)/N_t(i)\} \log[r_{ij}(\boldsymbol{\theta})/\{N_t(i,j)/N_t(i)\}]$$

$$\leq \frac{1}{2} \sum_{j\neq i} \{r_{ij}(\boldsymbol{\theta}) - N_t(i,j)/N_t(i)\}^2 .$$

Using this inequality in (17), one has

$$\sum_{i=1}^{M} N_t(i) \sum_{j\neq i} \{N_t(i,j)/N_t(i)\} \log\{N_t(i,j)/N_t(i)\}$$

$$\geq \sum_{i=1}^{M} \sum_{j\neq i} N_t(i, j) \log r_{ij}(\boldsymbol{\theta}) , \tag{18}$$

an inequality which is valid for all $\boldsymbol{\theta} \in \Theta$.

Now consider

$$\sum_{i=1}^{M} \{ N_t(i) \log q(i, \boldsymbol{\theta}) - A_t(i) q(i, \boldsymbol{\theta}) \}$$

$$= \sum_{i=1}^{M} N_t(i) \log [q(i, \boldsymbol{\theta})/\{N_t(i)/A_t(i)\}]$$

$$- \sum_{i=1}^{M} N_t(i) [q(i, \boldsymbol{\theta})/\{N_t(i)/A_t(i)\}]$$

$$+ \sum_{i=1}^{M} N_t(i) \log \{N_t(i)/A_t(i)\} ,$$

which by virtue of lemma 2 equals

$$- \sum_{i=1}^{M} N_t(i) - \sum_{i=1}^{M} N_t(i)[N_t(i)/A_t(i) - q(i, \boldsymbol{\theta})]^2 \{2z_i^2\}^{-1}$$

$$+ \sum_{i=1}^{M} N_t(i) \log[N_t(i)/A_t(i)]$$

where z_i lies between $N_t(i)/A_t(i)$ and $q(i,\boldsymbol{\theta})$. Thus for all $\boldsymbol{\theta} \in \Theta$,

$$\sum_{i=1}^{M} N_t(i) \log[N_t(i)/A_t(i)] - \sum_{i=1}^{M} N_t(i)$$

$$\geq \sum_{i=1}^{M} \{N_t(i) \log q(i,\boldsymbol{\theta}) - A_t(i)\, q(i,\boldsymbol{\theta})\} . \qquad (19)$$

Combine the inequalities (18) and (19), replace $\boldsymbol{\theta}$ by any c-approximate ML estimator $\hat{\boldsymbol{\theta}}_c(t)$ and divide throughout by N(t), to obtain the inequality

$$[N(t)]^{-1} [\sum_{i=1}^{M} \sum_{j \neq i} N_t(i,j) \log \{N_t(i,j)/N_t(i)\}$$

$$+ \sum_{i=1}^{M} N_t(i) \log \{N_t(i)/A_t(i)\} - \sum_{i=1}^{M} N_t(i)]$$

$$\geq N(t)^{-1} \{ \ell_t(\hat{\boldsymbol{\theta}}_c(t)) - \log P_{X(0)}(\boldsymbol{\theta})\}$$

$$\geq [N(t)]^{-1} \ell_t(\hat{\boldsymbol{\theta}}_c(t)) , \qquad (20)$$

which holds with probability one. Now, by definition of a c-approximate ML estimator,

$$L(V_t, \hat{\boldsymbol{\theta}}_c(t)) \geq c \sup [L(V_t, \boldsymbol{\theta}) \mid \boldsymbol{\theta} \varepsilon \Theta]$$

$$\geq c L(V_t, \boldsymbol{\theta}^0)$$

Thus

$$[N(t)]^{-1} [\sum_{i=1}^{M} \sum_{j \neq i} N_t(i,j) \log \{N_t(i,j)/N_t(i)\}$$

$$+ \sum_{i=1}^{M} N_t(i) \log \{N_t(i)/A_t(i)\} - \sum_{i=1}^{M} N_t(i)]$$

$$\geq [N(t)]^{-1} \ell_t(\hat{\boldsymbol{\theta}}_c(t))$$

$$\geq \{\log c + \log p_{X(0)}(\boldsymbol{\theta}^0)\}/N(t)$$

$$+\{ \sum_{i=1}^{M} \sum_{j \neq i} N_t(i,j) \log \{(r_{ij}(\boldsymbol{\theta}^0)\}/N(t)$$

$$+\{ \sum_{i=1}^{M} N_t(i) \log q(i,\boldsymbol{\theta}) - A_t(i) q(i,\boldsymbol{\theta}^0)\}/N(t). \quad (21)$$

Observe that by virtue of lemma 3, the lower and upper bounds for $[N(t)]^{-1} \ell_t(\hat{\theta}_c(t))$ provided by inequality (21), both converge, as $t \to \infty$, to the same finite limit with probability one. It follows that, as $t \to \infty$,

$$[N(t)]^{-1} [\ell_t(\hat{\boldsymbol{\theta}}_c(t)) - \sum_{i=1}^{M} \sum_{j \neq i} N_t(i,j) \log \{N_t(i,j)/N_t(i)\}$$

$$- \sum_{i=1}^{M} N_t(i) \log\{N_t(i)/A_t(i)\} + \sum_{i=1}^{M} N_t(i)]$$

$$\to 0$$

with $P_{\boldsymbol{\theta}^o}$-probability one. In particular, using expression (16) for $\ell_t(\boldsymbol{\theta})$, with $\boldsymbol{\theta}$ replaced by $\hat{\boldsymbol{\theta}}_c(t)$, it follows that

$$[N(t)]^{-1} [\sum_{i=1}^{M} \sum_{j\neq i} N_t(i,j) \log r_{ij}(\hat{\boldsymbol{\theta}}_c(t))/\{N_t(i,j)/N_t(i)\}]$$

$$+ \sum_{i=1}^{M} N_t(i) \log[q(i, \hat{\boldsymbol{\theta}}_c(t))/\{N_t(i)/A_t(i)\}]$$

$$- A_t(i)\, q(i, \hat{\boldsymbol{\theta}}_c(t) + \Sigma N_t(i)] \to 0 \tag{22}$$

almost surely $(P_{\boldsymbol{\theta}^o})$.

It is easy to check, by using lemma 1, that the first term in (22) is non-positive and by using lemma 2 that

$$\sum_{i=1}^{M} N_t(i) \log[q(i, \hat{\boldsymbol{\theta}}_c(t))/\{N_t(i)/A_t(i)\}]$$

$$- \sum_{i=1}^{M} A_t(i)\, q(i, \hat{\boldsymbol{\theta}}_c(t) + \Sigma N_t(i) \tag{23}$$

is also non-positive. Thus, as $t \to \infty$,

$$[N(t)]^{-1}\Sigma\, \Sigma\, N_t(i,j) \log[r_{ij}(\hat{\boldsymbol{\theta}}_c(t))/\{N_t(i,j)/N_t(i)\}] \tag{24}$$

as well as

$$[N(t)]^{-1} [\sum_{i=1}^{M} N_t(i) \log[q(i, \hat{\boldsymbol{\theta}}_c(t))/\{N_t(i)/A_t(i)\}]$$

$$- \sum_{i=1}^{M} A_t(i)\, q(i, \hat{\boldsymbol{\theta}}_c(t)) + \sum_{i=1}^{M} N_t(i)] , \tag{25}$$

both converge to zero with probability one. Apply lemma 1 to expression (24) to claim that

$$\Sigma\,[r_{ij}(\hat{\boldsymbol\theta}_c(t)) - N_t(i,j)/N_t(i)]^2 \to 0 \tag{26}$$

with probability one. Using lemma 2 for expression (25), we find that

$$\sum_{i=1}^{M} [N_t(i)/N(t)]\{\, q(i, \hat{\boldsymbol\theta}_c(t)) - N_t(i)/A_t(i)\}^2\,(2z_i^2)^{-1} \to 0 \tag{27}$$

almost surely. One can use results (6), (7) and (11) of lemma 3 to assert that, as $t \to \infty$,

$$\frac{N_t(i,j)}{N_t(i)} = \frac{N_t(i,j)}{t}\;\frac{t}{A_t(i)}\;\frac{A_t(i)}{N_t(i)}$$

$$\to\; q(i,\, j,\, \boldsymbol\theta^0)/q(i,\, \boldsymbol\theta^0) = r_{ij}(\boldsymbol\theta^0)$$

almost surely. Thus (26) and (27) imply that as $t \to \infty$,

$$r_{ij}(\hat{\boldsymbol\theta}_c(t)) \to r_{ij}(\boldsymbol\theta^0)$$

$$q(i,\, \hat{\boldsymbol\theta}_c(t)) \to q(i,\, \boldsymbol\theta^0)$$

with probability one. The identifiability condition immediately implies that $\hat{\boldsymbol\theta}_c(t) \to \boldsymbol\theta^0$ almost surely $(P_{\boldsymbol\theta^0})$. The proof is complete.

We have already pointed out that a c-approximate ML estimator is not unique. The above theorem asserts that every c-approximate ML estimator is strongly consistent. It is desirable to know whether they are uniformly consistent, i.e., whether for every fixed $c \in (0,1)$ and $\delta > 0$ one can find a $t_0(\omega,\, c,\, \delta)$ depending only on δ, c and the sample path ω, such that for all $t \geq t_0$, every c-approximate ML estimator belongs to the sphere $S(\boldsymbol{\theta}^o, \delta) = \{ \boldsymbol{\theta} \mid |\boldsymbol{\theta} - \boldsymbol{\theta}^o| \leq \delta\}$.

If this property holds, we can assert that for almost every realization of the finite Markov process of sufficiently long duration, whatever may be the c-approximate ML estimator we choose, it would be close to the true parametric point. This is essentially the assertion of the following

<u>Theorem 2</u> : Under the conditions of theorem 1, a c-approximate ML estimator is uniformly strongly consistent.

<u>Proof</u> : The inequality (21) provides upper and lower bounds for $[N(t)]^{-1} \ell_t(\hat{\boldsymbol{\theta}}_c(t))$ and these bounds are free of $\hat{\boldsymbol{\theta}}_c(t)$. They therefore hold for all $\hat{\boldsymbol{\theta}}_c(t)$. It therefore follows from (22) that

$$\inf_{\hat{\boldsymbol{\theta}}_c(t)} [N(t)]^{-1} \Big[\sum_{i=1}^{M} \sum_{j \neq i} N_t(i,j) \log[r_{ij}(\hat{\boldsymbol{\theta}}_c(t)/\{N_t(i,j)/N_t(i)\}]$$

$$+ \sum_{i=1}^{M} N_t(i) \log[q(i, \hat{\boldsymbol{\theta}}_c(t))/\{N_t(i)/A_t(i)\}]$$

$$- A_t(i)\, q(i, \hat{\boldsymbol{\theta}}_c(t)) + \Sigma N_t(i)] \Big] \to 0$$

almost surely. The argument after (22) and lemmas 1 and 2 can be used to demonstrate that with probability one

$$\sup_{\hat{\boldsymbol{\theta}}_c(t)} |r_{ij}(\hat{\boldsymbol{\theta}}_c(t)) - r_{ij}(\boldsymbol{\theta}^o)| \to 0 \tag{28}$$

and

$$\sup_{\hat{\boldsymbol{\theta}}_c(t)} \sum_{i=1}^{M} \{N_t(i)/N(t)\}\{q(i, \hat{\boldsymbol{\theta}}_c(t)) - N_t(i)/A_t(i)\}^2 (2z_i^2)^{-1} \to 0.$$

Here z_i lies between $q(i, \hat{\theta}_c(t))$ and $N_t(i)/N(t)$. Since

$$N_t(i)/N(t) = \frac{N_t(i)}{A_t(i)} \frac{A_t(i)}{t} \frac{t}{N(t)}$$

$$\rightarrow q(i,\theta^0)/\{ \sum_{r=1}^{M} \sum_{s \neq r} \pi_r(\theta^0)\, q(r,s,\theta^0)\}$$

almost surely, it follows that $q(i, \hat{\theta}_c(t))$ is uniformly bounded in $\hat{\theta}_c(t)$ for almost all realizations and all sufficiently large t. Furthermore $N_t(i)/N(t)$ is also almost surely uniformly bounded for variations in $\hat{\theta}_c(t)$. It follows that there exist constants μ_i such that $z_i \leq \mu_i$ with probability one for all sufficiently large t. Hence,

$$\sup_{\hat{\theta}_c(t)} \sum \frac{N_t(i)}{N(t)} \{ q(i, \hat{\theta}_c(t)) - \frac{N_t(i)}{A_t(i)} \}^2 \frac{1}{2\mu_i^2} \xrightarrow{a.s.} 0 .$$

In particular, for each $i \in S$

$$\sup_{\hat{\theta}_c(t)} | q(i, \hat{\theta}_c(t) - N_t(i)/A_t(i) | \xrightarrow{a.s.} 0$$

and since $N_t(i)/A_t(i) \xrightarrow{a.s.} q(i,\theta^0)$

$$\sup_{\hat{\theta}_c(t)} | q(i, \hat{\theta}_c(t)) - q(i, \theta^0) | \xrightarrow{a.s.} 0 . \qquad (29)$$

The relations (28) and (29) coupled with the identifiability condition imply that the c-approximate ML estimators are uniformly strongly consistent.

Although the theorems 1 and 2 establish the uniform strong consistency of the c-approximate ML estimators, they do not throw any light on the existence or otherwise of the ML estimators. A set of sufficient conditions for the existence of the ML estimators of $\boldsymbol{\theta}$ is provided by the following :

Theorem 1 : If the identifiability condition is satisfied and if, for $\boldsymbol{\theta} \in \Theta$, each $q(i,j,\boldsymbol{\theta})$ is a continuous function of $\boldsymbol{\theta}$, then for all sufficiently large t, an ML estimator of $\boldsymbol{\theta}$ exists with probability one. This ML estimator is strongly consistent.

Proof : The expression (16) for $\ell_t(\boldsymbol{\theta})$ combined with inequalities (18) and (19) which are valid for all $\boldsymbol{\theta} \in \Theta$, and the results of lemma 3 imply that with probability one

$$\sup_{\boldsymbol{\theta} \in \Theta} L(V_t, \boldsymbol{\theta}) = \sup_{|\boldsymbol{\theta}-\boldsymbol{\theta}^o| \leq \varepsilon} L(V_t, \boldsymbol{\theta}) \tag{30}$$

for any given $\varepsilon > 0$, provided t is sufficiently large. Since $L(V_t, \boldsymbol{\theta})$ is a continuous function of $\boldsymbol{\theta}$ by virtue of the assumption, it follows that the supremum in (30) is in fact attained inside the sphere $\{\boldsymbol{\theta} \,|\, |\boldsymbol{\theta} - \boldsymbol{\theta}^o| \leq \varepsilon\}$. Thus an ML estimator exists and is also obviously strongly consistent.

6. ML ESTIMATION : PARAMETRIC CASE

We have shown in the previous section that when the intensity rates depend on a vector parameter $\boldsymbol{\theta} = (\theta_1, \ldots, \theta_k) \in \Theta$ which is an open subset of the k-domensional Euclidean space $\mathbb{R}^k$, the c-approximate ML estimators are, under certain conditions, strongly consistent and that when $q(i, j, \boldsymbol{\theta})$ are continuous functions of $\boldsymbol{\theta}$, an ML estimator exists. These results however do not provide a method of obtaining an ML estimator of $\boldsymbol{\theta}$ when it exists. In this section, we investigate the properties of estimators obtained as a solution of the likelihood equation

$$\partial \ell_t(\boldsymbol{\theta})/\partial\theta_u = 0 , \qquad u = 1, \ldots, k , \tag{1}$$

where the log-likelihood $\ell_t(\boldsymbol{\theta})$ of $\boldsymbol{\theta}$ is assumed to satisfy certain assumptions to be made precise below. We essentially simplify the methods of Billingsley (1961). Specifically, the following assumptions are made.

A.1 For each $\boldsymbol{\theta} \in \Theta$, the finite Markov process is irreducible i.e., $q(i, \boldsymbol{\theta}) > 0$ for all $i \in S$. The set

$$D = \{(i, j) \mid i \neq j , i, j \in S , q(i,j,\boldsymbol{\theta}) > 0\}$$

does not depend on $\boldsymbol{\theta}$.

A.2 Every intensity rate $q(i, j, \boldsymbol{\theta})$ has continuous bounded derivatives upto third order throughout Θ.

A.3 The k x k matrix $\Sigma(\boldsymbol{\theta}) = ((\sigma_{uv}(\theta)))$, defined by

$$\sigma_{uv}(\boldsymbol{\theta}) = \sum_{(i,j)\in D} \{\pi_i(\boldsymbol{\theta})/q(i, j, \boldsymbol{\theta})\} \frac{\partial q(i,j,\boldsymbol{\theta})}{\partial\theta_u} \frac{\partial q(i,j, \theta)}{\partial\theta_v} ,$$

where $\{\pi_i(\boldsymbol{\theta}), i \in S\}$ is the unique invariant distribution of the Markov process, is non-singular for every $\boldsymbol{\theta} \in \Theta$.

Recall that the log-likelihood function is

$$\log p(X(0), \boldsymbol{\theta}) + \sum_{(i,j)\in D} N_t(i,j) \log q(i,j, \theta) - \sum_{i\in S} A_t(i)\, q(i,\boldsymbol{\theta})$$

where $p(i, \theta) = \Pr[X(0) = i]$ defines the initial distribution which may depend on θ. Since there is only one observation on the initial distribution and since we are interested mainly in the asymptotic results, we can and shall ignore the term $\log p(X(0),\boldsymbol{\theta})$ in all later calculations. Define, therefore

$$\ell_t(\boldsymbol{\theta}) = \sum_{(i,j)\varepsilon D} \{N_t(i,j) \log q(i,j,\boldsymbol{\theta}) - A_t(i)\, q(i,j,\boldsymbol{\theta})\} \qquad (2)$$

to be the log-likelihood function. Let $\hat{\boldsymbol{\theta}}(t) = (\hat{\theta}_1(t),\ldots,\hat{\theta}_k(t))$ be a solution of the likelihood equations

$$\partial\ell_t(\boldsymbol{\theta})/\partial\theta_u = 0 \;, \quad u = 1, \ldots, k \;. \qquad (3)$$

Lemma 1 : Suppose A.1, A.2 and A.3 hold. Then for all $\boldsymbol{\theta} \in \Theta$, as $t \to \infty$

(i) $t^{-1}\, \partial\ell_t(\boldsymbol{\theta})/\partial\theta_u \overset{a.s.}{\to} 0 \;, \quad u = 1, \ldots, k,$

(ii) $t^{-1}\, \partial^2\ell_t(\boldsymbol{\theta})/\partial\theta_u\partial\theta_v \overset{a.s.}{\to} \sigma_{uv}(\boldsymbol{\theta}) \;, \quad u, v, \ldots, k,$

(iii) $t^{-1}|\partial^3\ell_t(\boldsymbol{\theta})/\partial\theta_u\, \partial\theta_v\, \partial\theta_w| \overset{a.s.}{\to} \mu(\boldsymbol{\theta}) \;,$

where $\overset{a.s.}{\to}$ denotes almost sure $P_{\boldsymbol{\theta}}$ - convergence and $\mu(\boldsymbol{\theta})$ is bounded on Θ.

Proof : Observe that

$$\lambda_u(\boldsymbol{\theta}) = \partial\ell_t/\partial\theta_u = \sum_{(i,j)\varepsilon D} \{N_t(i,j) - A_t(i)q(i,j,\boldsymbol{\theta})\}\partial \log q(i,j,\boldsymbol{\theta})/\partial\theta_u,$$

$$\lambda_{uv}(\boldsymbol{\theta}) = \frac{\partial^2\ell_t(\boldsymbol{\theta})}{\partial\theta_u\partial\theta_v} = \sum_{(i,j)\varepsilon D} A_t(i) \frac{\partial \log q(i,j,\boldsymbol{\theta})}{\partial\theta_u}\frac{\partial q(i,j,\boldsymbol{\theta})}{\partial\theta_v}$$

$$+ \sum_{(i,j)\varepsilon D} \{N_t(i,j) - A_t(i)q(i,j,\boldsymbol{\theta})\} \frac{\partial^2 \log q(i,j,\boldsymbol{\theta})}{\partial\theta_u\, \partial\theta_v}$$

and

$$\lambda_{uvw}(\boldsymbol{\theta}) = \frac{\partial^3 \ell_t(\boldsymbol{\theta})}{\partial\theta_u\, \partial\theta_v\, \partial\theta_w} = - \sum_{i,j,\varepsilon D} A_t(i)\Big\{ \frac{\partial \log q(i,j,\boldsymbol{\theta})}{\partial\theta_u} \frac{\partial^2 q(i,j,\boldsymbol{\theta})}{\partial\theta_v\, \partial\theta_w}$$

$$+ \frac{\partial q(i,j,\boldsymbol{\theta})}{\partial\theta_v} \frac{\partial^2 \log q(i,j,\boldsymbol{\theta})}{\partial\theta_u\, \partial\theta_w} + \frac{\partial q(i,j,\boldsymbol{\theta})}{\partial\theta_w} \frac{\partial^2 \log q(i,j,\boldsymbol{\theta})}{\partial\theta_u\, \partial\theta_v}$$

$$+ \sum_{i,j\varepsilon D} \{N_t(i,j) - q(i,j,\boldsymbol{\theta})\, A_t(i)\} \frac{\partial^3 \log q(i,j,\boldsymbol{\theta})}{\partial\theta_u\, \partial\theta_v\, \partial\theta_w} \quad .$$

The results (i), (ii) and (iii) are then immediate consequence of lemma 5.3.

Lemma 2 : Let $g = (g_1, \ldots, g_k)$ be a continuous function from the k-dimensional Euclidean space $\mathbb{R}^k$ to itself. Suppose that for every point

$$\mathbf{x} = (x_1, \ldots, x_k) \quad , \quad \text{such that} \quad |\mathbf{x}| = 1 \, ,$$

$$\sum_{u=1}^{k} x_u\, g_u(\mathbf{x}) < 0 \quad .$$

Then there exists a point $\hat{\mathbf{x}}$, $|\hat{\mathbf{x}}| < 1$, such that $g(\hat{\mathbf{x}}) = 0$.

Proof : See Aitchison and Silvey (1958).

Theorem 1 : Under assumptions A.1, A.2 and A.3 the equations (1) have for all sufficiently large t, a solution $\hat{\boldsymbol{\theta}}(t) \,\varepsilon\, \Theta$ with $P_{\boldsymbol{\theta}^o}$- probability one. This solution is strongly consistent, almost surely unique and provides a local maximum of the likelihood $\ell_t(\boldsymbol{\theta})$, with probability one.

Proof : Let the true parametric point $\boldsymbol{\theta}^o = (\theta_1^o, \ldots, \theta_k^o)$ be an interior point of Θ and N be a neighbourhood of $\boldsymbol{\theta}^o$ such that $N \subset \Theta$. For $\boldsymbol{\theta} \in N$ consider the following Taylor expansion of $\lambda_u(\boldsymbol{\theta})$,

$$\lambda_u(\boldsymbol{\theta}) = \lambda_u(\boldsymbol{\theta}^o) + \sum_{v=1}^{k} (\theta_v - \theta_v^o)\, \lambda_{uv}(\boldsymbol{\theta}^o)$$

$$+ \alpha|\boldsymbol{\theta} - \boldsymbol{\theta}^o|^2 \, G(V_t), \qquad (4)$$

where $|\boldsymbol{\theta} - \boldsymbol{\theta}^o|$ is the usual Euclidean distance, $|\alpha| \leq k^2/2$, and $G(V_t) = \sup_{\boldsymbol{\theta}' \in N} |\lambda_{uvw}(\boldsymbol{\theta}')|$. By Lemma 1,

$$t^{-1}\lambda_u(\boldsymbol{\theta}^o) \overset{a.s}{\to} 0, \quad t^{-1}\lambda_{uv}(\boldsymbol{\theta}^o) \overset{a.s}{\to} -\sigma_{uv}(\boldsymbol{\theta}^o) \qquad (5)$$

and $G(V_t)/t \overset{a.s.}{\to} \mu$, a non-negative real number. Moreover, observe that

$$E[\lambda_u(\boldsymbol{\theta}^o)] = 0,$$

$$E[-\lambda_{uv}(\boldsymbol{\theta}^o)] = E[\lambda_u(\boldsymbol{\theta}^o)\,\lambda_v(\boldsymbol{\theta}^o)]$$

by virtue of theorem 4.2.2 and theorem 4.2.3.

Thus, for each t, the matrix $((E(-\lambda_{uv}(\boldsymbol{\theta}^o))))$ is a variance-covariance matrix and therefore has to be positive semi-definite. Using corollary 4.2.2 one can also verify that

$$t^{-1} E\{-\lambda_{uv}(\boldsymbol{\theta}^o)\} \to \sigma_{uv}(\boldsymbol{\theta}^o), \quad u, v = 1, \ldots, k.$$

Thus the limit matrix, being the limit of positive semi-definite matrices, has to be positive semi-definite. By assumption A.3, it is non-singular and therefore $\Sigma(\boldsymbol{\theta}^o)$ is a positive definite matrix.

One can then find a positive number β such that for any vector $Z = (z_1, \ldots, z_k)^T$, $\sum_{u=1}^{k} z_u^2 = 1$, we must have

$$Z^T \ \Sigma(\boldsymbol{\theta}^o) Z \ \geq \ \beta . \tag{6}$$

Suppose now an $\varepsilon > 0$ is specified. Choose $\delta > 0$ such that

$$\delta < \varepsilon , \quad \{\boldsymbol{\theta} | |\boldsymbol{\theta} - \theta^o| \leq \delta \} \subset N, \ \delta < \beta/\{3k^3(\mu+1)\} , \tag{7}$$

where μ is defined immediately after (5) and β by (6). By virtue of (5), there exists a $P_{\boldsymbol{\theta}^o}$-null set Λ such that for every sample path $\omega \in \Lambda^c$, with an obvious change of notation,

$$|t^{-1} \lambda_u(\omega , \boldsymbol{\theta}^o)| < \delta^2$$

$$0 \leq t^{-1} G(V_t(\omega)) < \mu + 1$$

$$|t^{-1} \lambda_{uv}(\omega , \boldsymbol{\theta}^o) + \sigma_{uv}(\boldsymbol{\theta}^o)| < \delta$$

for all u, v = 1, ..., k. Thus for sample paths $\omega \in \Lambda^c$ and for $|\boldsymbol{\theta} - \boldsymbol{\theta}^o| \leq \delta$, by (4)

$$|t^{-1}\{ \lambda_u(\omega, \boldsymbol{\theta}^o) + \sum_{v=1}^{k} (\theta_v - \theta_v^o) \sigma_{uv}(\boldsymbol{\theta}^o)\} |$$

$$= |t^{-1}[\lambda_u(\omega, \boldsymbol{\theta}^o) + \sum_{v=1}^{k} (\theta_v - \theta_v^o) \{ \lambda_{uv}(\omega, \boldsymbol{\theta}^o) + \sigma_{uv}(\boldsymbol{\theta}^o)$$

$$+ \alpha |\boldsymbol{\theta} - \boldsymbol{\theta}_o|^2 G(V_t(\omega))] |$$

$$\leq |t^{-1} \lambda_u(\omega, \boldsymbol{\theta}^o)| + t^{-1} \sum_{v=1}^{k} |\theta_v - \theta_v^o| \; | \lambda_{uv}(\omega, \boldsymbol{\theta}^o) + \sigma_{uv}(\boldsymbol{\theta}^o)$$

$$+ \alpha t^{-1} | \boldsymbol{\theta} - \boldsymbol{\theta}_o|^2 \; |G(V_t(\omega))|$$

$$\leq \delta^2 + k \; |\boldsymbol{\theta} - \boldsymbol{\theta}^o|\delta + k^2 \; |\boldsymbol{\theta} - \boldsymbol{\theta}_o|^2 \; (\mu+1)/2$$

$$\leq \quad 3k^2\delta^2 \; (\mu+1) \; .$$

Now use the conditions in (7) to see that on the sphere $\{\boldsymbol{\theta}| \; |\boldsymbol{\theta} - \boldsymbol{\theta}^o| = \delta \}$, for every $\omega \in \Lambda^c$,

$$\sum_{u=1}^{k} (\theta_u - \theta_u^o) \; \lambda_u(\omega, \boldsymbol{\theta})$$

$$\leq \sum_{u,v=1}^{k} (\theta_u - \theta_u^o)(\theta_v - \theta_v^o) \; \sigma_{uv}(\theta^o) + 3k^2\delta^3(\mu+1)$$

$$\leq \quad -\beta \; |\boldsymbol{\theta} - \boldsymbol{\theta}^o| + 3k^2\delta^3(\mu+1)$$

$$= \quad -\beta\delta + 3k^2\delta^3(\mu+1)$$

$$< \quad 0.$$

A straight forward application of lemma 2 implies that for a sufficiently large t and $\omega \in \hat{\Lambda}^c$, there exists a point $\hat{\boldsymbol{\theta}}(t)$ inside the sphere $\{\boldsymbol{\theta}||\boldsymbol{\theta} - \boldsymbol{\theta}^o|<\delta\} \subset \{\boldsymbol{\theta}||\boldsymbol{\theta} - \boldsymbol{\theta}^o|< \varepsilon\}$ such that

$$t^{-1} \lambda_u(\omega, \hat{\boldsymbol{\theta}}) = 0 \; .$$

In other words, for almost all sample paths ω , and for all sufficiently large t, the likelihood equations (1) have a solution $\hat{\boldsymbol{\theta}}(t)$ inside the sphere $\{\boldsymbol{\theta} \mid |\boldsymbol{\theta} - \boldsymbol{\theta}^0| < \varepsilon\}$ for any specified $\varepsilon > 0$. We have thus demonstrated the almost sure existence of a strongly consistent solution of the likelihood equations for all sufficiently large t.

In order to demonstrate that this solution $\hat{\boldsymbol{\theta}}(t)$ of equations (1), provides at least a local maximum of $\ell_t(\boldsymbol{\theta}), \boldsymbol{\theta} \in \Theta$, consider the following Taylor expansion of the second partial derivative $\lambda_{uv}(\boldsymbol{\theta})$, $\boldsymbol{\theta} \in N$

$$\lambda_{uv}(\boldsymbol{\theta}) = \lambda_{uv}(\boldsymbol{\theta}^0) + \alpha\, |\boldsymbol{\theta} - \boldsymbol{\theta}^0| \; G(V_t) \tag{8}$$

with $|\alpha| \leq k$. As before choose δ such that $\{\boldsymbol{\theta} \,|\, |\boldsymbol{\theta} - \boldsymbol{\theta}^0| \leq \delta\} \subset N$. Let $((z_{uv}(\boldsymbol{\theta})))$ be a k-square, symmetric matrix such that

$$|z_{uv}(\boldsymbol{\theta}) - \sigma_{uv}(\boldsymbol{\theta}^0)| < \delta \{ 1 + k(\mu+1)\} \tag{9}$$

for all u, v = 1, ..., k, then $((z_{uv}))$ is a positive definite matrix. Once again by virtue of (5), there exists a $P_{\boldsymbol{\theta}^0}$-null set H, such that for every $\omega \in H^c$, and all sufficiently large t,

$$|t^{-1} \lambda_{uv}(\omega, \boldsymbol{\theta}^0) + \sigma_{uv}(\boldsymbol{\theta}^0)| < \delta \tag{10}$$

and $0 \leq G(\omega, V_t(\omega)) < \mu+1$. The relations (8) and (10) imply that for $|\boldsymbol{\theta} - \boldsymbol{\theta}^0| \leq \delta$ and $\omega \in H^c$,

$$|t^{-1} \lambda_{uv}(\omega, \boldsymbol{\theta}^0) + \sigma_{uv}(\boldsymbol{\theta}^0)| < \delta + \delta k(\mu+1) .$$

Thus by (9) for chosen δ and all sufficiently large t, the matrix $((t^{-1} \lambda_{uv}(\boldsymbol{\theta})))$ is negative definite for all $\boldsymbol{\theta}$ in the open sphere

$|\theta - \theta^0| < \delta$. Thus one has at most one solution $\hat{\theta}(t)$ of the likelihood equations such that $|\hat{\theta}(t) - \theta^0| < \delta$ and this solution provides a local maximum of the likelihood $\ell_t(\theta)$. The proof is complete.

Remark : The above theorem provides the existence of an essentially unique root of the ML equations. Although this root is strongly consistent, it guarantees only a local maximum of the likelihood and not the global maximum. We have also modified slightly our likelihood function. Thus it is not quite correct to call this root, an ML estimator. We shall therefore call it a ML equation estimator.

Example 1 : Consider the random walk with reflecting barriers [cf. example 2.5.1] which is a finite Markov process on the state - space $\{0, 1, ..., M\}$ and whose non-zero intensity rates are

$$q(i, i+1) = \lambda , \quad i = 0, 1, ..., M-1 ,$$

$$q(i, i-1) = \mu , \quad i = 1, ..., M .$$

Moreover, $q(0) = \lambda$, $q(M) = \mu$ and $q(i) = (\lambda+\mu)$, $i = 1,...,(M-1)$. Hence $\lambda,\mu \in (0,\infty)$ so that we have the two dimensional parameter $\theta = (\lambda,\mu)$ with $(0, \infty) \times (0, \infty)$ as the parameter space. The Markov process is obviously irreducible and the set

$$D = \{(i, j) \mid i \neq j, \ q(i,j) > 0, \ i, j \in S\}$$

$$= \bigcup_{i=0}^{M-1} (i, i+1) \cup \bigcup_{i=1}^{M} (i, i-1)$$

does not depend on (λ,μ). Thus A.1 is satisfied. The assumption A.2 is easily seen to be valid.

The invariant distribution of the process is

$$\pi_i = \alpha(\lambda/\mu)^i , \qquad i = 0, 1, \ldots, M ,$$

where $\alpha = \{\sum_{i=0}^{M} (\lambda/\mu)^i\}^{-1}$. An easy computation, together with the identification $\theta_1 = \lambda$, $\theta_2 = \mu$, yields the fact that

$$\sigma_{uv}(\boldsymbol{\theta}) = \sum_{(i,j)\in D} \{\pi(i,\boldsymbol{\theta})/q(i,j,\boldsymbol{\theta})\} \frac{\partial q(i,j,\boldsymbol{\theta})}{\partial\theta_u} \frac{\partial q(i,j,\boldsymbol{\theta})}{\partial\theta_v}$$

$$= \begin{cases} (1 - \pi_M)/\lambda & \text{if } u = v = 1 , \\ (1 - \pi_0)/\mu & \text{if } u = v = 2 , \\ 0 & \text{otherwise.} \end{cases}$$

The matrix $\Sigma(\boldsymbol{\theta}) = \mathrm{diag}\,(\,(1 - \pi_M)/\lambda ,\ (1 - \pi_0)/\mu)$ is therefore non-singular for every $\lambda > 0$, $\mu > 0$; and thus A.3 is satisfied.

The log-likelihood based on the complete observation of the process over $[0, t]$ is

$$\ell_t(\lambda,\mu) = \{\sum_{i=0}^{M-1} N_t(i, i+1)\} \log\lambda + \{\sum_{i=1}^{M} N_t(i, i-1)\} \log\mu$$

$$- \lambda \sum_{i=0}^{M-1} A_t(i) - \mu \sum_{i=1}^{M} A_t(i)$$

$$= R(t) \log\lambda + L(t) \log\mu - (t - A_t(M))\lambda - (t - A_t(0)\mu ,$$

where $R(t)$ is the number of jumps to the right and $L(t)$ is the number of jumps to the left. Hence the likelihood equations are

$$R(t)/\lambda - [t - A_t(M)] = 0 ,$$

and

$$L(t)/\mu - [t - A_t(0)] = 0 .$$

The ML equation estimators are thus, with probability one,

$$\hat{\lambda}(t) = R(t)/(t - A_t(M)), \quad \hat{\mu}(t) = L(t)/(t - A_t(0))$$

validly defined for all t so large that $R(t)$, $L(t)$, $t - A_t(M)$ and $t - A_t(0)$ are positive. It is easy to check that the matrix of second derivatives of $\ell_t(\lambda,\mu)$ with respect to λ and μ is $\text{dig}\,(-1/\lambda^2, -1/\mu^2)$ which is negative definite and therefore $\hat{\lambda}(t)$ and $\hat{\mu}(t)$ provide the maximum of $\ell_t(\lambda,\mu)$.

Example 2 : Consider the Holgate model of elephant herds [cf. Example 2.5.3] which is a finite irreducible Markov process on the state-space $\{1, \ldots, M\}$ with non-zero intensity rates specified by

$$q(i, i+1) = (M-i)\lambda, \quad i = 1, \ldots, M-1,$$

$$q(i, i-1) = (i-1)\mu, \quad i = 1, \ldots, M,$$

and

$$q(i) = (M-i)\lambda + (i-1)\mu, \quad i = 1, \ldots, M.$$

The parameter $\theta = (\lambda,\mu)$ is two dimensional with parameter space $(0, \infty) \times (0, \infty)$. One can easily verify as in Example 1 that the assumptions A.1, A.2 and A.3 are satisfied in this case also. Here

$$\begin{aligned}\ell_t(\lambda,\mu) &= \sum_{i=1}^{M-1} N_t(i,i+1) \log\{(M-i)\lambda\} + \sum_{i=2}^{M} N_t(i,i-1)\log\{(i-1)\mu\} \\ &\quad - \sum_{i=1}^{M} A_t(i)\{(M-i)\lambda + (i-1)\mu\} \\ &= R(t)\log\lambda + L(t)\log\mu \\ &\quad - \sum_{i=1}^{M-1} (m-i)\, A_t(i) - \sum_{i=2}^{M} (i-1)\, A_t(i) \\ &\quad + \sum_{i=1}^{M-1} N_t(i,i+1)\log(M-i) + \sum_{i=2}^{M} N_t(i,i-1)\log(i-1),\end{aligned}$$

where as before $R(t)$ and $L(t)$ are the number of jumps to the right and left respectively. An easy algebra yields

$$\hat{\lambda}(t) = R(t) / \{ \sum_{i=1}^{M-1} (M-i)\, A_t(i) \} ,$$

$$\hat{\mu}(t) = L(t) / \{ \sum_{i=2}^{M} (i-1)\, A_t(i) \},$$

as the ML equation estimators of λ and μ respectively. One should now verify that they do provide a maximum of $\ell_t(\lambda,\mu)$ by computing the matrix of its second derivatives.

7. ASYMPTOTIC DISTRIBUTION OF ML EQUATION ESTIMATORS

In this section we investigate the asymptotic distribution of the ML equation estimator $\hat{\theta}(t)$, whose existence is established in section 6. Define

$$\lambda(t) = t^{-1/2} (\lambda_1(\theta^o),\ldots, \lambda_k(\theta^o))^T$$

and

$$Y(t) = t^{1/2} ((\hat{\theta}_1(t) - \theta_1^o),\ldots, (\hat{\theta}_k(t) - \theta_k^o))^T , \tag{1}$$

where as before

$$\lambda_u(\theta^o) = [\partial \ell_t(\boldsymbol{\theta})/\partial \theta_u]_{\theta = \theta^o} , \quad u = 1,\ldots, k,$$

$\hat{\theta}_u(t)$ is the ML equation estimator of θ_u and $\boldsymbol{\theta}^o$ denotes the true parameter. Our main purpose in this section is to obtain the asymptotic distributions of $\lambda(t)$ and $Y(t)$ as $t \to \infty$. Towards this end we introduce the following notation.

If ξ_t and η_t are two random vectors of the same dimension, $\xi_t \sim \eta_t$ stands for the statement that the difference $\xi_t - \eta_t$ converges to zero in probability : $\xi_t - \eta_t \xrightarrow{p} 0$. As usual $\xrightarrow{\pounds}$ denotes convergence in law, $N(0, \Sigma(\theta))$ stands for the normal distribution with zero mean vector and dispersion matrix $\Sigma(\theta)$. We let χ_r^2 denote a r.v. which has chi-square distribution with r degrees of freedom.

<u>Lemma 1</u> : Suppose ξ_t and η_t are both k-dimensional random vectors, $t \in IR^+$. Suppose Z is a k-dimensional random vector such that as $t \to \infty$,

$$\xi_t \xrightarrow{\pounds} Z .$$

If either

$$|\xi_t - \eta_t| \leq \varepsilon_t |\xi_t| , \quad \varepsilon_t \sim 0 \tag{2}$$

or

$$|\xi_t - \eta_t| \leq \varepsilon_t' |\eta_t| , \quad \varepsilon_t' \sim 0 \tag{3}$$

then $\xi_t \sim \eta_t$ and $\eta_t \xrightarrow{\pounds} Z$.

<u>Proof</u> : Suppose $|\varepsilon_t'| < 1/2$ and observe that the elementary inequality

$$|\xi_t - \eta_t| \leq \varepsilon_t' |\eta_t| \leq \varepsilon_t' \, | |\xi_t| + |\xi_t - \eta_t| \, |$$

implies that

$$|\xi_t - \eta_t| \leq \varepsilon_t' |\xi_t| / (1 - \varepsilon_t') \leq 2\varepsilon_t' |\xi_t| .$$

Thus if (3) holds then (2) holds. It is therefore enough to prove that if $\varepsilon_t \sim 0$, $\varepsilon_t |\xi_t| \sim 0$.

Let F be the disuri ut on function of $|Z|$. Since $\xi_t \overset{\mathcal{L}}{\to} Z$, $|\xi_t| \overset{\mathcal{L}}{\to} |Z|$. Let $\delta_0 > 0$ and chosse δ such that $\pm\, \delta_0/\delta$ are points of continuity of F. Since

$$\Pr[\, \varepsilon_t |\xi_t| \geq \delta_0] \leq \Pr[\, \varepsilon_t > \delta] + \Pr[\, |\xi_t| \geq \delta_0/\delta\,] \quad ,$$

it follows that

$$\limsup_t \; \Pr[\, \varepsilon_t \, |\xi_t| \geq \delta_0] \leq 1 - F(\delta_0/\delta) + F(-\,\delta_0/\delta) \; .$$

Letting $\delta \to 0$ such that δ_0/δ continues to be a point of continuity of F, we find that $\varepsilon_t \; |\xi_t| \sim 0$. It follows that $\xi_t \sim \eta_t$ and therefore $\eta_t \overset{\mathcal{L}}{\to} Z$.

We now introduce the assumption

<u>A.4</u> The intensity matrix $Q(\boldsymbol{\theta}) = ((\; q(i, j, \boldsymbol{\theta})\;))$ has simple characteristic roots for all $\boldsymbol{\theta} \in \Theta$.

In the following theorem, we need that A.4 should hold only for $\boldsymbol{\theta} = \boldsymbol{\theta}^0$. However, since $\boldsymbol{\theta}^0$ is not known, we make the broader assumption A.4.

<u>Theorem 1</u> : Suppose that the finite Markov process satisfies the assumptions A.1 - A.4. Then

(i) $\lambda(t) \overset{\mathcal{L}}{\to} Z$,

(ii) $\lambda(t) \sim \Sigma(\boldsymbol{\theta}^0)\, Y(t)$,

(iii) $Y(t) \sim \Sigma^{-1}(\boldsymbol{\theta}^0)\, \lambda(t)$,

(iv) $Y(t) \overset{\mathcal{L}}{\to} Z^*$,

(v) $2\{\ell_t(\hat{\boldsymbol{\theta}}(t)) - \ell_t(\boldsymbol{\theta}^0)\} \sim Y^T(t)\, \Sigma(\boldsymbol{\theta}^0)\, Y(t)$

$$\sim \lambda^T(t) \;\; \Sigma^{-1}(\boldsymbol{\theta}^0)\, \lambda\,(t)\, ,$$

(vi) $2\{\ell_t(\hat{\boldsymbol\theta}(t)) - \ell_t(\boldsymbol\theta^o)\} \xrightarrow{\mathcal{L}} \chi^2_k$,

where Z has $N(0, \Sigma(\boldsymbol\theta^o))$ distribition and Z^* has $N(0,\Sigma^{-1}(\boldsymbol\theta^o))$ distribution.

Proof : Observe that

$$t^{-1/2}\lambda_u(\boldsymbol\theta^o)$$

$$= \sum_{i,j\varepsilon D} t^{-1/2}\{N_t(i,j)-A_t(i)q(i,j,\boldsymbol\theta^o)\} \frac{1}{q(i,j,\boldsymbol\theta^o)} \left|\frac{\partial q(i,j,\boldsymbol\theta)}{\partial\theta_u}\right|_{\theta^o} .$$

By theorem 4.2, the asymptotic joint distribution of the r.v.s

$$\xi_{ij}(t) = t^{-1/2}\{N_t(i,j) - A_t(i)q(i,j, \boldsymbol\theta^o)\},\ i,j\ \varepsilon\, D$$

is normal with zero means, variances $\pi_i(\boldsymbol\theta^o)\, q(i, j, \boldsymbol\theta^o)$ and zero correlations. It follows after some elementary calculations that every linear combination

$$\alpha^T \lambda(t) = \sum_{u=1}^{k} \alpha_u \lambda_u(\boldsymbol\theta^o)$$

of $\sqrt{t}\,\lambda_1(\theta^o)$, ..., $\sqrt{t}\,\lambda_k(\theta^o)$ has asymptotically a normal distribution with zero mean and variance $\alpha^T \Sigma(\boldsymbol\theta^o)\alpha$. A simple application of the Cramèr-Wold theorem [cf. Moran (1968) , p. 267] implies result (i).

In order to establish (ii) let N, a neighbourhood of $\boldsymbol\theta^o$ and G a r.v., be defined as in the proof of theorem 6.1. Using the expansion (6.4) of $\lambda_u(\boldsymbol\theta)$ and the fact that $\hat{\boldsymbol\theta}(t)\ \varepsilon\ N$ the likelihood equation (6.1) becomes

$$t^{-1/2}\ \lambda_u(\boldsymbol{\theta}^o) + \sum_{u=1}^{k} Y_u(t)\ \{t^{-1}\ \lambda_{uv}(\boldsymbol{\theta}^o)\}$$

$$+\ \alpha\ |\hat{\boldsymbol{\theta}}(t) - \boldsymbol{\theta}^o|\ |Y_u(t)|\ t^{-1}\ G(V_t) = 0$$

where $\alpha \leq k^2/2$. Hence the almost sure convergence results of lemma 6.1 imply that

$$|\lambda(t) - Y(t)| \leq \varepsilon(t)\ |Y(t)|$$

where $\varepsilon(t) \sim 0$. Lemma 1 now yields result (ii) as well as the fact that

$$\Sigma\ (\boldsymbol{\theta}^o)\ \ Y(t)\ \overset{\pounds}{\to}\ Z$$

where Z has $N(0, \Sigma(\boldsymbol{\theta}^o))$ distribution. But then (ii) implies (iii) and therefore

$$Y(t) \overset{\pounds}{\to} Z^*$$

which has $N(0, \Sigma^{-1}(\boldsymbol{\theta}^o))$ distribution. This proves result (iv).

The proof of results (v) and (vi) require us to consider the following Taylor expansion of $\ell_t(\boldsymbol{\theta})$ with $\boldsymbol{\theta} \in N$:

$$\ell_t(\boldsymbol{\theta}) = \ell_t(\boldsymbol{\theta}^o) + \sum_{u=1}^{k} (\theta_u - \theta_u^o)\ \lambda_u(\boldsymbol{\theta}^o)$$

$$+\ \frac{1}{2} \sum_{u=1}^{k} \sum_{v=1}^{k} (\theta_u - \theta_u^o)(\theta_v - \theta_v^o)\ \lambda_{uv}(\boldsymbol{\theta}^o)$$

$$+\ \alpha\ |\boldsymbol{\theta} - \boldsymbol{\theta}^o|^3\ \ G(V_t)\ ,$$

where now $\alpha \leq k^3/6$. Thus when $\hat{\boldsymbol{\theta}}(t) \in N$,

$$2\{\max_{\boldsymbol{\theta}\,\varepsilon\,\Theta} \ell_t(\boldsymbol{\theta}) - \ell_t(\boldsymbol{\theta}^o)\} = 2\{\ell_t(\hat{\boldsymbol{\theta}}) - \ell_t(\boldsymbol{\theta}^o)\}$$

$$= 2 \sum_{u=1}^{k} (\hat{\theta}_u(t) - \theta_u^o)\, \lambda_u(\boldsymbol{\theta}^o)\, \Sigma\, \Sigma(\hat{\theta}_u(t) - \theta_u^o)(\hat{\theta}_v(t) - \theta_v^o)\lambda_{uv}(\boldsymbol{\theta}^o)$$

$$+ \quad \alpha|\hat{\boldsymbol{\theta}}(t) - \boldsymbol{\theta}^o|^3 \quad G(V_t)\ . \tag{4}$$

Since, as $t \to \infty$,

$$|\,\hat{\boldsymbol{\theta}}(t) - \boldsymbol{\theta}^o|^3\ G(V_t) = |Y(t)|^3\ t^{-3/2}\ G(V_t) \ \overset{P}{\to}\ 0$$

we can use (4) to assert that

$$2\{\ell_t(\hat{\boldsymbol{\theta}}(t)) - \ell_t(\boldsymbol{\theta}^o)\} \sim 2 \sum_{u=1}^{k} t^{1/2}(\hat{\theta}_u(t) - \theta_u^o)\ t^{-1/2}\ \lambda_u(\boldsymbol{\theta}^o)$$

$$+ \Sigma\ \Sigma\ Y_u(t)\ Y_v(t)\ \{t^{-1}\ \lambda_{uv}(\boldsymbol{\theta}^o)\}\ .$$

Applying the convergence results (6.5) once again, we have

$$2[\ell_t(\hat{\boldsymbol{\theta}}(t) - \ell_t(\boldsymbol{\theta}^o)] \sim 2\ Y^T(t)\ \lambda(t) - Y^T(t)\ \Sigma(\boldsymbol{\theta}^o)\ Y(t)$$

$$\sim 2\ Y^T(t)\ \Sigma(\boldsymbol{\theta}^o)\ Y(t) - Y^T(t)\ \Sigma(\boldsymbol{\theta}^o)\ Y(t)$$

$$= Y^T(t)\ \Sigma(\boldsymbol{\theta}^o)\ Y(t) \tag{5}$$

$$\sim \lambda^T(t)\ \Sigma^{-1}(\boldsymbol{\theta}^o)\ \lambda(t)\ , \tag{6}$$

which is result (v).

Since $Y(t) \to Z^*$, which has $N(0, \Sigma^{-1}(\boldsymbol{\theta}^o))$ distribution, the result (vi) follows from standard properties of multivariate normal distribution [cf. Rao (1973), p. 188] .

The proof is complete.

Remark : A careful perusal of the proof of this theorem indicates that we have used assumption A.4 only to claim asymptotic normality of the r.v.s

$$\{N_t(i,j) - q(i,j)\, A_t(i)\} / \sqrt{t} \ .$$

Any other assumption which gurantees this result can also be substituted for A.4. We refer to Billingsley (1961) for a martingale approach to this problem.

Example 1 : The ML equation estimators for the parameters λ and μ of the random walk with reflecting barriers [cf. Examples 2.5.1 and 6.1] are

$$\hat{\lambda}(t) = R(t)/\{t - A_t(M)\} \quad \text{and} \quad \hat{\lambda}(t) = L(t)/\{t - A_t(0)\}$$

respectively. We have also seen in example 2.5.1 that the characteristic roots of the Q-matrix are distinct. Thus A.4 is also satisfied.

Observe that with $\boldsymbol{\theta} = (\lambda, \mu)$ and $\boldsymbol{\theta}^{o} = (\lambda_0, \mu_0)$

$$\Sigma(\boldsymbol{\theta}^{o}) = \text{diag}\ \{(1 - \pi_M)/\lambda_0 \ , \ (1 - \pi_0)/\mu_0\}$$

so that $\Sigma^{-1}(\boldsymbol{\theta}^{o}) = \text{diag}\ \{\lambda_0/(1 - \pi_M)\ , \ \mu_0/(1 - \pi_0)\}$ where λ_0, μ_0 are the true values of the parameters. It follows by theorem 1, that $\sqrt{t}\ \{\hat{\lambda}(t) - \lambda_0\}$ and $\sqrt{t}\{\hat{\mu}(t) - \mu_0\}$ are asymptotically independently and normally distributed with zero means and variances $\lambda_0/(1 - \pi_M)$ and $\mu_0/(1-\pi_0)$ respectively.

A similar argument establishes that for the Holgate model of elephant herds [cf. examples 2.5.3 and 6.2],

$$\sqrt{t}\ \{\hat{\lambda}(t) - \lambda_0\} \ , \quad \sqrt{t}\ \{\hat{\mu}(t) - \mu_0\}$$

are asymptotically independently and normally distributed with zero means and variances

$$\lambda_0/\{\sum_{i=1}^{M-1} (M - i)^2 \pi_i\} \quad \text{and} \quad \mu_0/\{\sum_{i=2}^{M} (i-1)^2 \pi_i\}$$

respectively.

The main result of theorem 1 is, of course, result (iii) which asserts that the asymptotic joint distribution of the ML equation estimators is the $N(0, \Sigma^{-1}(\boldsymbol{\theta}^o))$ distribution. We now make certain observations on the efficiency of the ML estimators of the intensity rates $q(i,j)$ as well as of the ML equation estimator $\hat{\boldsymbol{\theta}}(t)$ of θ.

Suppose $X_1, \ldots, X_n$ are i.i.d. r.v.s with common density function $f(x,\boldsymbol{\theta})$ depending on a vector-valued parameter $\theta=(\theta_1,\ldots,\theta_k)$ say.The k x k matrix $I(\boldsymbol{\theta}) = ((\, i_{uv}(\boldsymbol{\theta})\,))$ defined by

$$i_{uv}(\boldsymbol{\theta}) = E[\frac{\partial \log f(X_1,\boldsymbol{\theta})}{\partial\theta_u} \frac{\partial \log f(X_1, \boldsymbol{\theta})}{\partial\theta_v}], \quad u,v=1,\ldots,k$$

is called the information matrix per observation in view of the fact that

$$i_{uv}(\boldsymbol{\theta}) = \frac{1}{n} E[\frac{\partial \log L(X_1,\ldots, X_n, \theta)}{\partial\theta_u} \frac{\partial \log L(X_1,\ldots, X_n, \theta)}{\partial\theta_v}$$

where $L(X_1, \ldots, X_n, \boldsymbol{\theta})$ is the likelihood $\prod_{j=1}^{n} f(X_j, \theta)$ of θ based on the random sample $(X_1, \ldots, X_n)$ of size n. Accordingly we may define the matrix $I(\theta, t) = ((\, i_{uv}(\theta,t)))$ by

$$i_{uv}(\boldsymbol{\theta}, t) = \frac{1}{t} E [\frac{\partial \log L(V_t,\boldsymbol{\theta})}{\partial\theta_u} \frac{\partial \log L(V_t, \boldsymbol{\theta})}{\partial\theta_v}],$$

where $L(V_t, \boldsymbol{\theta})$ is the likelihood of $\boldsymbol{\theta}$ corresponding to the complete observation $V_t = \{X(u),\ 0 \le u \le t\}$ on the continuous parameter stochastic process. However, unlike the i.i.d. case, $I(\boldsymbol{\theta}, t)$ may not be free of t. We can then define the information matrix as

$$I(\theta) = \lim_{t\to\infty} I(\boldsymbol{\theta}, t)$$

provided the limit exists.

We shall say that an estimator $\xi(t)$ of θ based on V_t is asymptotically efficient if $\sqrt{t}(\xi(t) - \theta^o)$ has asymptotically the normal $N(0, I^{-1}(\boldsymbol{\theta}))$ distribution, where we assume that $I(\boldsymbol{\theta})$ is a non-singular matrix. Furthermore $\xi(t)$ is said to be first order efficient in the sense of Rao [cf. Rao (1973), p. 348] if there exists a non-random $k \times k$ matrix $B(\boldsymbol{\theta})$ possibly depending on $\boldsymbol{\theta}$, such that as $t \to \infty$

$$\sqrt{t}\ \{\xi(t) - B(\theta^o)\ \eta(t)\} \to 0$$

either in probability or with probability one, where

$$\eta(t) = \frac{1}{t}\left[\frac{\partial \ell_t(\boldsymbol{\theta})}{1}, \ldots, \frac{\partial \ell_t(\boldsymbol{\theta})}{u}\right]_{\boldsymbol{\theta} = \boldsymbol{\theta}^o}$$

$\ell_t(\boldsymbol{\theta})$ being $\log L(V_t, \boldsymbol{\theta})$.

<u>Theorem 2</u> : Under the assumption A.4, the ML estimators $q(i,j,t)$ of the intensity rates $q(i,j)$, $i, j \in D$, are asymptotically effi - cient and first order efficient.

<u>Proof</u> : According to corollary 4.1,

$$\sqrt{t}(\hat{q}(i,j,t) - q(i,j)) , \quad i, j \in D$$

are asymptotically independently and normally distributed with zero meansand variances $q(i,j)/\pi_i$. It is easy to check by virtue of Corollary 4.2.2 that

$$\lim_{t\to\infty} \frac{1}{t} E\left\{ \frac{\partial \ell_t}{\partial q(i,j)} \cdot \frac{\partial \ell_t}{\partial q(r,s)} \right\}$$

$$= \begin{cases} \pi_i/q(i,j) & , \text{ if } i = r, \ j = s \ , \\ 0 & , \text{ otherwise.} \end{cases}$$

Thus $\hat{q}(i,j,t)$ are all asymptotically efficient.

By employing lemma 5.3 and theorem 4.2 it is readily seen that for each $(i, j) \varepsilon D$

$$\sqrt{t}\left[\hat{q}(i,j,t) - q(i,j) - \frac{q(i,j)}{\pi_i} \frac{\partial \ell_t}{\partial q(i,j)}\right]$$

$$= \frac{t}{A_t(i)} \left(\frac{N_t(i,j) - q(i,j)\, A_t(i)}{t} \right) \left(1 - \frac{q(i,j)}{\pi_i} \frac{A_t(i)}{t} \right)$$

$$\overset{P}{\to} \ .$$

This establishes the first order efficiency of $\hat{q}(i,j,t)$ and the proof is complete.

<u>Theorem 3</u> : Under assumptions A.1 - A.4 , the ML equation estimator $\hat{\boldsymbol{\theta}}(t)$ of $\boldsymbol{\theta}$ is asymptotically efficient and first order efficient.

<u>Proof</u> : Since

$$\ell_t(\boldsymbol{\theta}) = \sum_{(i,j)\varepsilon D} \{ N_t(i,j) \log q(i,j,\theta) - A_t(i)\, q(i,j,\boldsymbol{\theta}) \} \ ,$$

it is easy to establish by using corollary 4.2.2 that

$$\frac{1}{t} E\left(\frac{\partial \ell_t(\boldsymbol{\theta})}{\partial \theta_u} \frac{\partial \ell_t(\boldsymbol{\theta})}{\partial \theta_v} \right) \to \sigma_{uv}(\boldsymbol{\theta})$$

as defined by assumption A.3. The asymptotic efficiency follows from result (iv) of theorem 1 and the first order efficiency is an immediate consequence of (iii) of the same theorem.

8. TESTS OF HYPOTHESES

We have investigated the properties of the ML estimators of the intensity rates and of the ML equation estimators of the parameters in the previous sections. We now develop certain likelihood based tests of hypotheses about finite irreducible Markov processes. As in the case of estimation, we shall be able to develop asymptotic tests based on large period of observation, in view of the intractability of the finite time properties of the test statistics.

Suppose $\{ X(t), t \in IR^{+}\}$ is a finite, irreducible Markov process governed by the intensity rates $q(i,j)$,$(i, j) \in D$ which is the set of pairs (i,j) of states for which $q(i,j) > 0$. Let the null hypothesis H_0 be the simple hypothesis

$$q(i,j) = q^{o}(i,j) \quad , \quad (i, j) \in D ,$$

where $q^{o}(i,j)$ are specified positive numbers. Let the alternative hypothesis H_1 be the negation of H_0; viz.,

$$q(i,j) \neq q^{o}(i,j) \quad , \quad (i, j) \in D.$$

We have already seen in section 3 that the ML estimator of $q(i,j)$ based on the complete observation $V_t = \{X(u), 0 \leq u \leq t\}$ of the process is

$$\hat{q}(i, j, t) = N_t(i,j)/A_t(i) \quad , \quad (i, j) \in D.$$

Then the log likelihood evaluated at $\hat{q}(i,j,t),(i, j) \in D$ is

$$\ell_t(\hat{q}(i,j,t)) = \sum_{i,j\varepsilon D} N_t(i,j) \log \{N_t(i,j)/A_t(i)\}$$

$$- A_t(i) \sum_{v\varepsilon D_i} N_t(i,v)/A_t(v)$$

where $D_i = \{v \mid (i, v) \varepsilon D\}$. It follows that the appropriate Neyman-Pearson statistic [cf. Rao (1973), p. 417] for testing H_0 against H_1 is

$$\Lambda(t) = 2\{\ell_t(\hat{q}(i,j,t) - \ell_t(q^0(i,j))\}$$

$$= \sum_{i,\, j\varepsilon D} [N_t(i,j) \log [N_t(i,j)/\{A_t(i)\, q^0(i,j)\}]$$

$$- A_t(i) \{N_t(i,j)/A_t(i) - q^0(i,j)\}] . \qquad (1)$$

The following theorem asserts that $\Lambda(t)$ has asymptotically a chi-square distribution.

Theorem 1 : The Neyman -Pearson statistic $\Lambda(t)$ defined by (1) is asymptotically equivalent to the chi-square statistic

$$\eta(t) = \Sigma\{N_t(i,j) - A_t(i)\, q^0(i,j)\}^2/\{A_t(i)\, q^0(i,j)\} \qquad (2)$$

in the sense that under the null hypothesis $H_0, \Lambda(t) - \eta(t) \overset{P}{\to} 0$ as $t \to \infty$. Moreover, if the intensity matrix $Q^0 = ((q^0(i,j)))$ has distinct characteristic roots, then $\eta(t)$ (and therefore $\Lambda(t)$) has, under H_0, asymptotically a chi-square distribution with d degrees of freedom where d is the number of elements in D.

Proof : In order to establish asymptotic equivalence of $\Lambda(t)$ and $\eta(t)$, it is enough to demonstrate that for each $i \varepsilon S$, the difference

$$2 \sum_{j \in D_i} [\, N_t(i,j) \log [N_t(i,j)/\{A_t(i)\, q^o(i,j)\}] - \{N_t(i,j) - A_t(i)\, q^o(i,j)\} \,] \qquad (3)$$

$$- 2 \sum_{j \in D_i} \{N_t(i,j) - A_t(i)\, q^o(i,j)\}^2/\{A_t(i)\, q^o(i,j)\}$$

converges in probability to zero as $t \to \infty$. Observe then that the first term in (3) is

$$2 \sum_{j \in D_i} A_t(i)\, q^o(i,j) \left[\frac{N_t(i,j)}{A_t(i)\, q^o(i,j)} \log \left\{ \frac{N_t(i,j)}{A_t(i)\, q^o(i,j)} \right\} - \left\{ \frac{N_t(i,j)}{A_t(i)\, q^o(i,j)} - 1 \right\} \right] . \qquad (4)$$

By virtue of lemma 5.3, under H_0,

$$N_t(i,j) /\{A_t(i)\, q^o(i,j)\} \overset{a.s.}{\to} 1, \quad \text{as } t \to \infty .$$

Thus for all sufficiently large t and almost all ω,

$$|1 - N_t(i,j) /\{A_t(i)\, q^o(i,j)\} | < 1/2 .$$

Write $Y(t) = A_t(i)\, q^o(i,j)$ and

$$Z(t) = 1 - N_t(i,j)/\{A_t(i)\, q^o(i,j)\} .$$

It follows that for all sufficiently large t and almost surely,

$$\log [N_t(i,j)/\{A_t(i)\, q^o(i,j)\}] = \log \{1 - z(t)\}$$

$$= - \{Z(t) + Z^2(t)/2 + \theta\; Z^3(t)\}$$

where $|\theta| \leq 1$. The expression (4) becomes

$$2 \sum_{j\varepsilon D_i} Y(t)[\{1 - Z(t)\}\{- Z(t) - Z^2(t)/2 - \theta Z^3(t)\} + Z(t)]$$

$$= 2 \sum_{j\varepsilon D_i} Y(t)\{Z^2(t)/2 - (\theta-1) Z^3(t) - \theta Z^4(t)\}$$

$$= \sum_{j\varepsilon D_i} \{N_t(i, j) - A_t(i) q^o(i,j)\}^2/\{A_t(i) q^o(i,j)\}$$

$$- 2 \sum_{j\varepsilon D_i} Y(t)\{-(\theta-1) Z^3(t) - \theta Z^4(t)\} .$$

It follows that with probability one and for all sufficiently large t,

$$|\Lambda(t) - \eta(t)| \leq 2 \sum_{i,\ j\varepsilon D} |Y(t)|\{|\theta - 1||Z^3(t)| + \theta|Z^4(t)|\}.$$

Observe now that

$$|Y(t)\ Z^3(t)| = \{t^{-1}A_t(i)q^o(i,j)\}^{-2}\{t^{-2/3}|N_t(i,j) - A_t(i)q^o(i,j)|\}^3$$

in which the first factor converges almost surely to $\{q^o(i)\}^{-2}$ and the second factor, which equals

$$t^{-1/2}[|N_t(i,j) - A_t(i)\ q^o(i,j)|/\sqrt{t}\,]^3$$

is easily seen to converge in probability to zero by virtue of theorem 4.2. This and a similar argument implies that, as $t \to \infty$,

$$|Y(t)|\ |Z^r(t)| \xrightarrow{P} 0, \quad r = 3,4 .$$

Thus $$|\Lambda(t) - \eta(t)| \xrightarrow{P} 0 .$$

The statistic $\eta(t)$ is easily seen to be the sum of squares of d asymptotically independent standard normal variates by virtue of theorem 4.2. It follows that $\eta(t)$ and therefore $\Lambda(t)$ has under H_0 asymptotically a chi-square distribution with d degrees of freedom. The proof is complete.

The above result can be used to test the null hypothesis H_0 against the alternative H_1. The usual procedure is to reject H_0 in favour of H_1 for large values of $\Lambda(t)$ or equivalently of $\eta(t)$. In particular, we reject H_0 at level α , $0 < \alpha < 1$, whenever $\Lambda(t)$ exceeds $\chi_d^2(\alpha)$, where $\chi_d^2(\alpha)$ is the $100(1-\alpha)\%$ point of the chi-square distribution defined by

$$\Pr[\ \chi_d^2 > \chi_d^2(\alpha)\ ] = \alpha \quad .$$

Example 1 : Consider the random walk $\{X(t)\ ,\ t \in \mathbb{R}^+\}$ on the state-space $\{0, 1, \ldots, M\}$ with reflecting barriers at 0 and M. Let H_0 be the simple null hypothesis

$$q(i, i+1) = \lambda_i = \lambda^o, \quad i = 0, \ldots, M-1$$

$$q(i, i-1) = \mu_i = \mu^o, \quad i = 1, \ldots, M,$$

all other $q(i,j)$ being zero and let H_1 be the composite hypothesis that $q(i,i+1) \neq \lambda^o, i = 0, \ldots, M-1$ and $q(i,i-1) \neq \mu^o, i=1,\ldots,M$ are the only positive intensity rates .

The ML estimators of $q(i, i+1)$ and $q(i, i-1)$ are $N_t(i, i+1) / A_t(i)$ and $N_t(i, i-1)/A_t(i)$ respectively. The appropriate test statistic is

$$\Lambda_1(t) = 2[\, \ell_t(\hat{q}(i,j,t)) - \ell_t(\lambda^o, \mu^o)]$$

$$= 2 \sum_{i=0}^{M-1} [N_t(i,i+1) \log [N_t(i,i+1)/\{A_t(i)\lambda^o\}]$$

$$- \{N_t(i, i+1) - A_t(i)\, \lambda^o \}\,]$$

$$+ 2 \sum_{i=1}^{M} N_t(i,i-1) \log[N_t(\bar{i},i-1)/\{A_t(i)\mu^o\}]$$

$$- \{N_t(i, i-1) - A_t(i)\mu^o \}\,]$$

which has asymptotically a chi-square distribution with 2M degrees of freedom. The test statistic $\Lambda(t)$ is asymptotically equivalent to the following chi-square statistic :

$$\sum_{i=0}^{M-1} [N_t(i, i+1) - A_t(i)\, \lambda^o\,]^2/\{A_t(i)\, \lambda^o\, \}$$

$$+ \sum_{i=1}^{M} [N_t(i, i-1) - A_t(i)\, \mu^o]^2/\{A_t(i)\mu^o\}\ ,$$

and the later can be used in place of $\Lambda(t)$.

Now suppose that the intensity rates of the irreducible finite Markov process depend on a vector-valued parameter $\theta = (\theta_1, \ldots, \theta_k) \in \mathbb{R}^k$ and that assumptions A.1-A.3 hold. Let H_0 be the simple null hypothesis $\boldsymbol{\theta} = \boldsymbol{\theta}^o$, which is to be tested against the alternative hypothesis H_1 : $\boldsymbol{\theta} \neq \boldsymbol{\theta}^o$ at level α. Suppose further that the intensity matrix $Q(\boldsymbol{\theta}^o)$ at $\boldsymbol{\theta} = \boldsymbol{\theta}^o$ has distinct characteristic roots. Then according to theorem 6.1, there exists a consistent root $\hat{\boldsymbol{\theta}}(t)$ of the likelihood equation. Moreover, according to result (vi) of theorem 7.1 , the Neyman-Pearson statistic $\Lambda(t) = 2\, \{\ell_t(\hat{\boldsymbol{\theta}}(t)) - \ell_t(\boldsymbol{\theta}^o)\}$ has asymptotically a chi-square distribution with k degrees of freedom. Then as in case of the intensity rates, we carry out the test for H_0 against H_1 by rejecting H_0 in favour of H_1 whenever $\Lambda(t) > \chi_k^2(\alpha)$. It is also possible to adopt the arguments of theorem 1 above and show that in the parametric case also, $\Lambda(t)$ is equivalent to the chi-square statistic

$$\eta(t) = \sum_{(i,j)\in D} \{N_t(i,j) - A_t(i)\, q(i,j,\theta^o)\}^2/\{A_t(i)\, q(i,j,\theta^o)\}.$$

Example 2 : In the random walk with reflecting barriers, suppose we wish to test the simple null hypothesis $\lambda = \lambda^o$, $\mu = \mu^o$ against the alternative that $\lambda \neq \lambda^o$, and/or $\mu \neq \mu^o$. The appropriate test statistic is

$$\Lambda_2(t) = 2\,[\ell_t(\hat{\lambda}(t)), \hat{\mu}(t) - \ell_t(\lambda^o, \mu^o)]$$

$$= 2 \sum_{i=0}^{M-1} [N_t(i,i+1) \log [\hat{\lambda}(t)/\lambda^o] - \{N_t(i,i+1) - A_t(i)\lambda^o\}]$$

$$+ 2 \sum_{i=1}^{M} [N_t(i,i-1) \log[\hat{\mu}(t)/\mu^o] - \{N_t(i,i-1) - A_t(i)\mu^o\}]$$

$$= 2\, R(t) \log \{\hat{\lambda}(t)/\lambda^o\} - 2[R(T) - \{t - A_t(M)\}\, \lambda^o]$$

$$+ 2L(t) \log \{\hat{\mu}(t)/\mu^o\} - 2[L(t) - \{1 - A_t(0)\}\, \mu^o]$$

which has asymptotically a chi-square distribution with 2 degrees of freedom.

Both the cases discussed above deal with the tests of simple null hypotheses against composite alternative hypotheses. The case when the null hypothesis is also a composite hypothesis, requires some additional preparation.

Suppose, as before, that the intensity rates of the irreducible finite Markov process depend on a k-dimensional parameter $\boldsymbol{\theta} = (\theta_1, \ldots, \theta_k) \in \Theta \subset \mathbb{R}^k$, and that assumptions A.1 - A.4 hold. Let Φ be an open subset of $\mathbb{R}^r$, $r < k$ and let h be a function from Φ to Θ, i.e. if $\varphi = (\varphi_1, \ldots, \varphi_r) \in \Phi$, then $h(\varphi) = (h_1(\varphi), \ldots, h_k(\varphi)) \in \Theta$. We are interested in testing the composite null hypothesis that the intensity rates are $q(i,j, h(\varphi))$, $\varphi \in \Phi$, against the alternative that they are $q(i,j,\theta)$, $\theta \in \Theta$. The set Φ and the function $h(\varphi)$ are not arbitrary. They are assumed to satisfy the following assumptions.

<u>B.1</u> The co-ordinate functions $h_u(\varphi)$ of $h(\varphi)$ have continuous third order partial derivatives and the $k \times r$ matrix

$$K(\varphi) = ((\, k_{uv}(\varphi) \,)) = ((\, \partial h_u(\varphi)/\partial \varphi_v \,)) ,$$

$u = 1, \ldots, k$, $v = 1, \ldots, r$ has rank r throughout Φ.

<u>B.2</u> The intensity rates $q(i, j, h(\varphi))$ determined by φ satisfy A.1 for all $\varphi \in \Phi$, i.e., the function h is such that the finite Markov process continues to be irreducible for all $\varphi \in \Phi$ and that the set

$$\begin{aligned} \{(i,j) \mid i \neq j \,,\, q(i,j, h(\varphi)) > 0 \} &= \{(i,j) \mid i \neq j \quad q(i, j,\boldsymbol{\theta}) > 0 \} \\ &= D , \end{aligned}$$

is the same for all $\boldsymbol{\theta} \in \Theta$ and $\varphi \in \Phi$.

By virtue of B.1 and B.2, it follows that the intensity rates $q(i,j, h(\varphi))$, as functions defined on Φ, have continuous partial derivatives of order three.

The rxr matrix $\Sigma^*(\varphi) = ((\sigma^*_{uv}(\varphi)))$ corresponding to the matrix $\Sigma(\boldsymbol{\theta})$ of A.3 is defined by the relation

$$\sigma^*_{uv}(\varphi) = \sum_{(i,j)\varepsilon D} \frac{\pi_i(h(\varphi))}{q(i,j,h(\varphi))} \frac{\partial q(i,j, h(\varphi))}{\partial \varphi_u} \frac{\partial q(i,j, h(\varphi))}{\partial \varphi_v} .$$

Since

$$\frac{\partial q(i,j,h(\varphi))}{\partial \varphi_u} = \sum_{r=1}^{k} \frac{\partial q(i,j,h(\varphi))}{\partial h_r(\varphi)} \frac{\partial h_r(\varphi)}{\partial \varphi_u} ,$$

one can easily see that

$$\Sigma^*(\varphi) = k^T(\varphi) \Sigma(h(\varphi)) k(\varphi)$$

and since Σ is non-singular for all $\boldsymbol{\theta} \varepsilon \Theta$, and rank of $k(\varphi)$ is r, $\Sigma^*(\varphi)$ is also non-singular for all $\varphi \varepsilon \Phi$.

One consequence of the discussion so far is that the intensity rates $q(i,j, h(\varphi))$ as functions of $\varphi \varepsilon \Phi$ satisfy the analogues of the assumptions A.1 - A.3. This implies that the likelihood equations

$$\partial \ell_t(h(\varphi))/\partial \varphi_u = 0, \quad u = 1, \ldots, r$$

have, for all sufficiently large t and almost surely a solution $\hat{\varphi}(t)$, say, which is strongly consistent and which almost surely provides a local maximum of $\ell_t(h(\varphi))$.

Before proceeding any further with the problems of tests of hypotheses, we digress slightly and establish the following theorem which is in the nature of an Analysis of Variance theorem. Its corollary is of interest to us in our later discussion.

Although this theorem is established by Billingsley (1961), our proof is more statistical in nature.

Theorem 2 : Let X be a $p\times 1$ random vector with normal $N(0, \Sigma)$ distribution, Σ being a p-square positive definite matrix. Let $r_1, \ldots, r_\nu$ be ν positive integers such that $1 \leq r_1 < \ldots < r_\nu = p$. Define H_k to be a $r_{k+1} \times r_k$ matrix with real elements and let the rank of H_k be r_k, $k = 1, \ldots, \nu-1$. Let $M_\nu = I_p$ and $M_k = H_{\nu-1} H_{\nu-2} \ldots H_k$, $k = 1, \ldots, \nu-1$. Then

(i) $M_k^T \Sigma M_k$ is a non-singular matrix,

(ii) $Q_k = X^T M_k (M_k^T \Sigma M_k)^{-1} M_k^T X$, $k = 1, \ldots, \nu - 1$,

has a chi-square distribution with r_k degrees of freedom ; and

(iii) the ν quadratic forms $Q_1, Q_2 - Q_1, \ldots, Q_\nu - Q_{\nu-1}$ are independently distributed, the distribution of $Q_{k+1} - Q_k$, being a chi-square distribution with $r_{k+1} - r_k$ degrees of freedom.

Proof : The result (i) is trivially true for $k = \nu$, since $M_\nu^T \Sigma M_\nu = \Sigma$ which is non-singular by assumption. Suppose (i) holds for $k = \nu - j$, $0 \leq j < \nu$. Consider

$$M^T_{\nu-j-1} \Sigma M_{\nu-j-1} = H^T_{\nu-j-1} (M^T_{\nu-j} \Sigma M_{\nu-j}) H_{\nu-j-1}$$

and observe that Σ being positive definite, $M^T_{\nu-j} \Sigma M_{\nu-j}$ is a Gramian matrix [cf. Rao (1973), p. 69] of rank $r_{\nu-j}$ by the induction hypothesis. Hence there exists a $r_{\nu-j}$-square matrix G of rank $r_{\nu-j}$ such that

$$M_{\nu-j}^T \Sigma M_{\nu-j} = G^T G .$$

It follows that

$$M_{\nu-j-1}^T \Sigma M_{\nu-j-1} = (H_{\nu-j-1}^T G^T)(G H_{\nu-j-1})$$

and therefore has rank $r_{\nu-j-1}$. The usual induction argument yields (i).

Since Σ is a positive definite matrix, there exists a symmetric matrix D such that $D\Sigma D = I_p$. Then the random vector $Y = DX$ has $N(0, I_p)$ distribution. Define

$$K_{\nu-1} = D^{-1} H_{\nu-1} , \quad K_k = H_k ,$$

$$N_\nu = D^{-1} , \quad N_k = K_{\nu-1} \cdots K_k , \; k = 1, \ldots, \nu-2$$

It is easily verified that

$$M_k^T \Sigma M_k = N_k^T N_k$$

and by virtue of (i) rank $(N_k^T N_k)$ is r_k; $k = 1, \ldots, \nu$.

Now $M_k = DN_k$, and therefore

$$\begin{aligned} Q_k &= X^T D^T N_k (N_k^T D \Sigma D N_k)^{-1} N_k^T D X \\ &= Y^T N_k (N_k^T N_k)^{-1} N_k^T Y \\ &= Y^T A_k Y , \text{ say ,} \end{aligned}$$

where $A_k = N_k (N_k^T N_k)^{-1} N_k^T$ has rank r_k. It is easy to check that $A_k = A_k^2$, so that A_k is an idempotent matrix and therefore Q_k has chi-square distribution with r_k degrees of freedom; [cf. Rao (1973), p. 186]. This proves (ii).

In order to establish (iii), let $j < k$, and observe that

$$N_j = N_{j+1} K_j = N_k K_{k-1} K_{k-2} \cdots K_j .$$

Therefore

$$A_k A_j = N_k (N_k^T N_k)^{-1} N_k^T N_k K_{k-1} \cdots K_j (N_j^T N_j)^{-1} N_j^T$$

$$= N_k K_{k-1} \cdots K_j (N_j^T N_j)^{-1} N_j^T$$

$$= N_j (N_j^T N_j)^{-1} N_j^T = A_j .$$

Similarly one can show that $A_j A_k = A_j$. It follows that for $j < k$,

$$(A_{k+1} - A_k)(A_{j+1} - A_j) = 0 \qquad (5)$$

Thus

$$Q = Y^T Y = \sum_{k=1}^{\nu-1} (Q_{k+1} - Q_k) + Q_1$$

together with the relation (5) implies (iii) by virtue of the Fisher-Cochran theorem and its modifications [cf. Rao (1973), p. 165] .

Now suppose $\{\xi(t),\ t \in \mathbb{R}^+\}$ is a vector-valued stochastic process such that as $t \to \infty$, $\xi(t) \overset{\mathcal{L}}{\to} X$ of the above lemma. Then using the fact that if g is a continuous non-random function, then $g(\xi(t)) \overset{\mathcal{L}}{\to} g(X)$ [cf. Rao (1973), p. 124] we can assert the following

<u>Corollary</u> : Suppose the assumptions of theorem 2 hold and that the random $p \times 1$ vector $\xi(t) \overset{\mathcal{L}}{\to} X$. Then as $t \to \infty$

$$Q_k(t) = [\xi(t)]^T M_k (M_k^T \Sigma M_k)^{-1} M_k^T \xi(t)$$

$$\overset{\pounds}{\rightarrow} \chi^2_{r_k}$$

and the $\nu-1$ r.v.s $Q_{k+1}(t) - Q_k(t)$, $k = 1, \ldots, \nu-1$ are asymptotically independent.

Now let φ^o be the true value of the parameter i.e., we assume that the intensity rates governing the Markov process are $q(i,j,\boldsymbol{\theta}^o)$, $(i, j) \in D$, where $\boldsymbol{\theta}^o = h(\varphi^o)$. Define the random vector

$$\lambda^*(t) = t^{-1/2} \{\lambda_1(h(\varphi^o)), \ldots, \lambda_r(h(\varphi^o))\}^T ,$$

where

$$\lambda_u(h(\varphi^o)) = [\partial \ell_t(h(\varphi^o))/\partial \varphi_u]_{\varphi = \varphi^o}$$

$$= \left[\sum_{j=1}^{r} \frac{\partial \ell_t(h(\varphi))}{\partial h_j(\varphi)} \frac{\partial h_j(\varphi)}{\partial \varphi_u} \right]_{\varphi = \varphi^o} .$$

Then $\lambda^*(t) = [K(\varphi^o)]^T \lambda(t)$, where $\lambda(t)$ is defined by (7.1).

Let us take $\nu = 2$, $r_1 = r$, $r_2 = k$, H_1 to be the kxr matrix $K(\varphi)$ which is of rank r, and $M_2 = I_r$ in theorem 2 above. Then by a straight forward application of its corollary, we have the following theorem in which we write $\ell_t^*(\varphi)$ for $\ell_t(h(\varphi))$.

<u>Theorem 3</u> : Under the assumptions B.1, B.2, A.3 and A.4 as $t \to \infty$,

$$\text{(i)} \quad \Lambda_1(t) = 2\{ \ell_t(\hat{\boldsymbol{\theta}}(t)) - \ell_t(\boldsymbol{\theta}^o)\} \overset{\pounds}{\rightarrow} \chi^2_k$$

$$\text{(ii)} \quad \Lambda_2(t) = 2\{ \ell_t(\hat{\varphi}(t)) - \ell_t^*(\varphi^o)\} \overset{\pounds}{\rightarrow} \chi^2_{k-r}$$

$$\text{(iii)}\ \Lambda_3(t) = 2\{\ \ell_t(\hat{\boldsymbol{\theta}}(t)) - \ell_t^*(\hat{\varphi}(t))\} \overset{\mathcal{L}}{\to} \chi^2_{k-r}$$

and $\Lambda_2(t)$ and $\Lambda_3(t)$ are asymptotically independent.

The results of this theorem can be used to construct appropriate tests of hypotheses. Thus, for example $\Lambda_1(t)$ and $\Lambda_2(t)$ are the test statistics for testing the hypotheses $\boldsymbol{\theta} = \boldsymbol{\theta}^o$ and $\varphi = \varphi^o$ respectively against their negations as the alternatives. The statistic $\Lambda_3(t)$ is the test statistic for the null hypothesis that the intensity rates are $q(i,j,\varphi)$, $\varphi \varepsilon \Phi$ against the alternative that they are $q(i,j,\boldsymbol{\theta}),\boldsymbol{\theta} \varepsilon \Theta$. In all these tests the null hypothesis is rejected for large values of the test statistic.

<u>Example 3</u> : Suppose that for the random walk with reflecting barriers, we wish to test the composite null hypothesis

$$\begin{aligned} H^* \ :\ & q(0,1) = q(1,2) = \ldots = q(M-1\ ,\ M) > 0\ , \\ & q(M,M-1) = q(M-1,\ M-2) = \ldots = q(1,0) > 0\ , \\ & q(i,j) = 0\ ,\quad j \neq i \neq 1\ ,\quad i = 0,\ 1,\ldots,\ M, \end{aligned}$$

against the composite alternative hypothesis

$$\begin{aligned} H^{**} \ :\ & q(i,\ i+1) > 0\ ,\quad i = 0,\ 1,\ \ldots,\ M-1, \\ & q(i,\ i-1) > 0\ ,\quad i = 1,\ \ldots,\ M, \\ & q(i,\ j) = 0\ ,\quad j \neq i \pm 1\ ,\quad i = 0,\ 1,\ldots,\ M. \end{aligned}$$

Here the parameter space Θ is the positive quadrant of the 2M-dimensional Euclidean space and Φ is the positive quadrant of the 2-dimensional Euclidean space. Thus the (r+1)-th co-ordinate θ_{r+1} of $\boldsymbol{\theta}$ is $q(r,\ r+1)$ for $0 \leq r \leq M-1$ and the (M+r+1)-th co-ordinate θ_{M+1+r} is $q(r,\ r-1)$, $r = 1,\ldots,M$. Let λ and μ

denote common values of $q(i, i+1)$, $i = 0, 1, \ldots, M-1$ and $q(i, i-1)$, $i = 1, \ldots, M$ respectively as specified by H^*.

The h function from Φ to Θ is

$$h(\lambda,\mu) = (\lambda, \ldots, \lambda, \mu, \ldots \mu) .$$

If λ is identified with φ_1 and μ with φ_2, then the matrix $K(\lambda,\mu)$ is

$$\begin{pmatrix} E_{1M} & 0_{1M} \\ 0_{1M} & E_{1M} \end{pmatrix},$$

which is obviously of rank 2. The other assumptions are also easily verified.

The appropriate Neyman-Pearson test statistic for testing H^* against H^{**} is

$$\begin{aligned} \Lambda_3(t) &= 2\{ \ell_t(q(i, j, t)) - \ell_t(\hat{\lambda}(t), \hat{\mu}(t))\} \\ &= 2 \sum_{i=0}^{M-1} [N_t(i,i+1) \log [N_t(i,i+1)/\{A_t(i) \hat{\lambda}(t)\}] \\ &\qquad - \{N_t(i,i+1) - A_t(i) \hat{\lambda}(t) \}] \\ &\quad + 2 \sum_{i=1}^{M} [N(i, i-1) \log [N_t(i,i-1)/\{A_t(i) \hat{\mu}(t)\}] \\ &\qquad - \{N_t(i,i-1) - A_t(i) \hat{\mu}(t)\}], \end{aligned}$$

which has asymptotically a chi-square distribution with 2M-2 degrees of freedom. By theorem 3, the test statistics $\Lambda_2(t)$ of Example 2 and $\Lambda_3(t)$ of this example are asymptotically independent.

9. CONCLUDING REMARKS

The results of the previous sections clearly bring out that we have a fairly satisfactory asymptotic theory of estimation and of testing of hypotheses based on a single realization from a finite irreducible Markov process. It should be evident that in this development, irreducibility of the process, which guarantees a steady limit behaviour, is a crucial assumption for the asymptotic theory.

In our studies, we have implicitely assumed that the initial distribution and the number of states are known. The assumption about the initial distribution is made because in a single realization, we have only one observation on the initial distribution and therefore only the trivial estimator $\delta_{X(0)i}$ of $p_i(0)$, $i \varepsilon S$. In case $p_i(0)$ depends on a parameter $\boldsymbol{\theta}$, which also defines the intensity rates, then one can use its ML estimator to estimate the initial distribution as well as test hypotheses about it.

The assumption that the number of states in the finite Markov process is known is usually valid. However, in some radiobiological situations, the total number of states is unknown. Bharucha - Reid (1956) has suggested an ML estimator for the number of states of a finite Markov chain with two absorbing states. His method can be used for a Markov process also. However, if the finite Markov process is irreducible then a natural estimator is the number of distinct states visited by the process. It can be easily seen to be strongly consistent.

Let us now examine the effect of dropping the assumption of irreducibility of the process. Suppose then that in the M-state process, states 1, ..., m, $m < M$, form a minimal closed class and that the remaining M-m states m+1, ..., M, form the class T of transient states. If the initial state $X(0) \varepsilon C$, the process will, forever remain in C. The asymptotic theory developed in this chapter is then applicable to the Markov process on C. If all the intensity rates $q(i,j)$, $i \neq j$, i, j = 1,..., M, depend on the same parameter θ, the realization, restricted to states in C, will provide adequate information on $\boldsymbol{\theta}$ and one can carry out statistical inference for the entire Markov process. On the other hand, suppose all the intensity rates are unknown or that the intensity rates $q(i,j)$, $i \neq j$, i,j = 1,..., m, depend on the parameter $\boldsymbol{\theta}$ and that the intensity rates corresponding to the transient states depend on parameter φ, which is functionally independent of $\boldsymbol{\theta}$. If $X(0) \varepsilon C$, we shall have no information on $q(i,j)$, i, $j \varepsilon T$ or on φ. Even if $X(0) \varepsilon T$, since absorption in C occurs in finite time with probability one, there is no asymptotic theory available for inference on φ. It is thus evident that a single realization is quite inadequate for inference relating to transient states.

The situation becomes more complex if there are two or more minimal closed classes and a class of transient states. Let C_1 and C_2 be two minimal closed classes and let T denote the class of transient states which lead to states in C_1 and in C_2. The difficulties mentioned in the previous paragraph continue to exist. However, consider the simple situation when all the intensity rates depend on a single parameter $\boldsymbol{\theta}$. The results of theorem 7.1 imply that $\sqrt{t}(\hat{\boldsymbol{\theta}}(t) - \boldsymbol{\theta}^o)$, has asymptotically as $t \to \infty$, the $N(0,\Sigma_1(\boldsymbol{\theta}^o))$ distribution if observations are restricted to C_1 and has $N(0, \Sigma_2(\boldsymbol{\theta}^o))$ distribution if they are restricted to C_2. Suppose $X(0) = j \varepsilon T$ and that $\alpha_{j1}(\boldsymbol{\theta}^o)$ and $\alpha_{j2}(\boldsymbol{\theta}^o) = 1 - \alpha_{j1}(\boldsymbol{\theta}^o)$ are the probabilities of absorption in the closed classes C_1 and C_2 respectively when $j \in T$ is the initial state. It is easy to verify that as $t \to \infty$, $\sqrt{t}(\hat{\theta}(t) - \theta)$ converges in distribution to a r.v. Z which has distribution function

$$\alpha_{j1}(\boldsymbol{\theta}^o)\, \Phi_1(x) + \alpha_{j2}(\boldsymbol{\theta}^o)\, \Phi_2(x) ,$$

where $\Phi_r(x)$ is the distribution function of the $N(0, \Sigma_r(\boldsymbol{\theta}^o))$ distribution, $r = 1,2$. Thus the asymptotic distribution of the estimators $\hat{\boldsymbol{\theta}}(t)$ is no longer normal but is a mixture of two normal distributions. The theory developed in the previous section is therefore no longer applicable. However, if one is willing to carry out conditional inference, conditional on absorption in C_1 or C_2 as the case may be, the previous asymptotic theory is applicable.

It should now be evident that when one is dealing with a reducible Markov process, it is best to observe a large number of independent realizations of finite durations, whenever such a scheme is feasible.

It is rather unfortunate that no finite time (small sample) theory of inference is available for finite Markov processes. The major difficulty is of course the non-availability of tractable results for the distribution of the sufficient statistics, $\{N_t(i,j),\ A_t(i),\ i \neq j,\ i, j \in S\}$. Recent results of Adke and Manjunath (1984) and Manjunath (1984) indicate that it is preferable to use sequential methods of estimation. They demonstrate that certain functions of the intensity rates/parameters can be estimated efficiently' (in the sense of unbiasedness and minimum variance) provided the process is observed over a random duration which is a stopping time.

PROBLEMS

1. Let $\{Y_n, n \in Z^+\}$ be a sequence of independent and identically distributed discrete r.v.s and let $\{N(t), t \in \mathbb{R}^+\}$ be an independent Poisson process. Define $S_n = Y_0 + Y_1 + \ldots + Y_n$, $n \in Z^+$ and let $X(t) = S_{N(t)}$, $t \in \mathbb{R}^+$. Show that $\{X(t), t \in \mathbb{R}^+\}$ is a discrete Markov process. Obtain the transition probabilities of the $X(t)$ - process when

(i) $Pr[Y_n = 1] = p$, $Pr[Y_n = 0] = 1-p$, $0 < p < 1$,

(ii) $Pr[Y_n = k] = \exp(-\lambda)\lambda^k/k!$, $k \in Z^+$, $\lambda > 0$.

(iii) $Pr[Y_n = 1] = p$, $Pr[Y_n = -1] = 1-p$, $0 < p < 1$.

Classify the states of the discrete Markov process in each of the above cases.

2. Let $\{X_n, n \in Z^+\}$ be a sequence of r.v.s such that $\{X_{3n+1}, X_{3n+2}, n \in Z^+\}$ are i.i.d. r.v.s with

$$Pr[X_{3n+1} = 1] = 1/2 = Pr[X_{3n+1} = 0]$$
$$= Pr[X_{3n+2} = 1] = Pr[X_{3n+2} = 0].$$

Let $X_{3n+3} = |X_{3n+1} - X_{3n+2}|$, $n \in Z^+$. Suppose $\{N(t), t \in \mathbb{R}^+\}$ is a Poisson process, independent of the sequence $\{X_n, n \in Z^+\}$. Define $\{X(t) = X_{N(t)}, t \in \mathbb{R}^+\}$. Show that $\{X(t), t \in \mathbb{R}^+\}$ is not a Markov process although, $p_{ij}(t) = Pr[X(t) = j | X(0) = i]$, $i,j = 0,1$, satisfy the Chapman - Kolmogorov equations.

3. Let

$$Q_1 = \begin{pmatrix} -1 & 1 & 0 \\ 0 & -1 & 1 \\ 1 & 0 & -1 \end{pmatrix} \quad \text{and} \quad Q_2 = \begin{pmatrix} -1 & 1/2 & 1/2 \\ 1/2 & -1 & 1/2 \\ 1/2 & 1/2 & -1 \end{pmatrix}$$

be the intensity matrices of two Markov processes. Obtain their transition probabilities and show that their discrete skeletons to scale $h = 4\pi/\sqrt{3}$ have the same family of finite dimensional distributions.

Speakman (1967)

4. Let $\{X(t),\ t \in \mathbb{R}^+\}$ be a finite irreducible Markov process with transition probability matrix $P(t)$. Let $P = P(1)$. A finite Markov chain with transition matrix P^* is said to be <u>imbeddable</u> in a finite Markov process with the same state-space iff there exists a finite Markov process such that $P(1) = P^*$.

(i) Show that if a stochastic matrix P^* can be imbedded in a finite Markov process, then the determinant $|P^*|$ of P^* is positive.

(ii) If $|P^*| > 0$, then show that P^* is imbeddable in a finite Markov process iff there exists an intensity matrix Q^* such that $P^* = \exp(Q^*)$. Such a P^* is called a skeleton.

(iii) Show that a 2x2 stochastic matrix P^* is a skeleton iff $|P^*| > 0$ implies and is implied by the fact that trace $(P^*) > 1$.

Kingman (1962)

5. Show that for an irreducible finite Markov process there exists a $\lambda > 0$ such that for all states i, j

$$t^{-1} \log p_{ij}(t) \to -\lambda$$

as $t \to \infty$. Show further that there exist constants α_{ij} such that

(i) $p_{ij}(t) \leq \alpha_{ij} \exp(-\lambda t)$, $i \neq j$,

(ii) $p_{ii}(t) \leq \exp(-\lambda t)$,

(iii) $\lambda \leq \min \{q_i \mid i \in S\}$.

Kingman (1963)

6. Let $\{X(t), t \in \mathbb{R}^+\}$ be finite irreducible birth-death process with state-space $\{0, 1, \ldots, M\}$, positive birth rates λ_n , $n = 0, 1, \ldots, M-1$ and positive death rates μ_n , $n = 1, \ldots, M$. Define two sequences of polynomials as follows :

$$P_0 = 1, \quad P_1 = s + \lambda_0$$

$$P_r = (s + \lambda_{r-1} + \mu_{r-1}) P_{r-1} - \lambda_{r-2} \mu_{r-1} P_{r-2} , \quad 2 \leq r \leq M+1,$$

and

$$Q_0 = 1, \quad Q_1 = s + \mu_M$$

$$Q_r = (s + \lambda_{M+1-r} + \mu_{M+1-r}) Q_{r-1} - \lambda_{M+1-r} \mu_{M+2-r} Q_{r-2} ,$$

$$2 \leq r \leq M+1 .$$

Let $c_{jj} = 1$ $0 \leq j \leq M$, and

$$c_{j,k} = \begin{cases} \mu_{k+1} \cdots \mu_j \, , & k < j \, , \\ \lambda_j \cdots \lambda_{k-1} & k > j \, . \end{cases}$$

Further let β_r be defined by

$$P_{M+1}(s) = Q_{M+1}(s) = s(s + \beta_1) \ldots (s + \beta_M) \, ,$$

$0 < \beta_1 < \ldots < \beta_M$. Put

$$a_{j,k,r} = \frac{c_{jM} \, c_{Mk} \, P_j(-\beta_r) \, P_k(-\beta_r)}{P_M \, (-\beta_r) \, P'_{M+1}(-\beta_r)} \, .$$

Then prove that

$$p_{jk}(t) = \pi_k + \sum_{r=1}^{M} a_{n,j,r} \exp(-\beta_r t) \, ,$$

where $\pi_k = \lim_{t \to \infty} p_{jk}(t)$.

Rosenlund (1978)

7. Suppose $\{X(t), t \in \mathbb{R}^+\}$ is an irreducible birth-death process with state-space $\{0, 1, \ldots, M\}$ birth rates $\lambda_0, \ldots, \lambda_{M-1}$ and death rates $\mu_1, \ldots, \mu_M$.

(i) Show that the intensity matrix Q of such a process has distinct characteristic roots.

(ii) If $\eta(x) = \{\eta_0(x), \ldots, \eta_M(x)\}$ is the right characteristic vector of Q corresponding to the characteristic root $(-x)$, show that $\eta_j(x)$ is a real polynomial of degree j, such that

$$\sum_{i=0}^{M} \eta_i(x_\alpha)\, \eta_i(x_\beta)\, \pi_i = \delta_{\alpha\beta}/\rho_\alpha ,$$

where (a) $-x_\alpha$ and $-x_\beta$ are the characteristic roots of Q,

(b) $\pi_i = \{\lambda_0 \lambda_1 \cdots \lambda_{i-1}\}/\{\mu_1 \mu_2 \cdots \mu_i\}$, $i = 0,1,\ldots,M$,

(c) $\rho_\alpha = \sum_{i=0}^{M} \eta_i^2(x_\alpha)\pi_i > 0$, $\alpha,\beta = 0, 1, \ldots, M$.

(iii) Show that the transition probabilities are

$$p_{ij}(t) = \pi_j \sum_{\alpha=0}^{M} \exp(-x_\alpha t)\, \eta_i(x_\alpha)\, \eta_j(x_\alpha)\, \rho_\alpha .$$

Karlin and McGregor (1957)

8. Two urns, urn I and urn II, contain in all N balls. At the epochs of occurrence of a Poisson event with parameter $N\mu$, a ball is selected at random out of N balls and is placed in urn I with probability p and in urn II with probability $1-p$, $0 < p < 1$. If $X(t)$ is the number of balls in urn I at epoch t, show that $\{X(t), t \in \mathbb{R}^+\}$ is a finite Markov process. Obtain its transition probabilities, classify its states and find the invariant distribution, if any.

Karlin and McGregor (1965)

9. Let $\{X(t), t \in \mathbb{R}^+\}$ be a finite, irreducible birth - death process on the set $\{0, 1, \ldots, M\}$. Use the forward equations satisfied by its transition probabilities to obtain a system of linear equations for the Laplace transform

$$p_{jk}^{*}(\theta) = \int_0^{\infty} e^{-\theta t} p_{jk}(t)dt, \quad \theta > 0 ,$$

of $p_{jk}(t)$. Show that the $(M+1)$ equations have a unique solution and use Cramer's rule to obtain the Laplace transform $p_{jk}^{*}(\theta)$. Hence or otherwise obtain the Laplace transform of first passage density functions.

Moran (1963)

10. Let $\{X(t), t \in IR^+\}$ be a two-state Markov process with state-space $\{0, 1\}$ and intensity rates λ and μ. Let $A_t(0)$ be the total time spent in state 0 and $N(t)$ the total number of changes of states in $(0, t)$. Obtain the moment generating function of $A_t(0)$ and the probability generating function of $N(t)$.

Pedler (1971)

11. Let $\{X(t), t \in IR^+\}$ be an irreducible finite Markov process. In the terminology introduced in section 4.3, let Y_r denote the time to r-th occurrence of an A-transition, where A is a proper subset of the set of all possible direct transitions $(i \to j)$, $i, j \in S$. Obtain the moment generating functions (m.g.f.) of Y_r, $Y_{r+s} - Y_r$ and the joint m.g.f. of Y_r and Y_{r+s}.

Darroch and Morris (1967)

12. Suppose $\{X(t), t \in IR^+\}$ is a finite Markov process with transient states $1,\dots, m$ and absorbing states $m+1,\dots, M$.

Obtain the variance of the duration $A_t(j)$ of stay in state j, $1 \leq j \leq M$ during $(0, t)$ given that the initial state is i, $1 \leq i \leq m$. Obtain a closed form solution when the $m \times m$ sub - matrix of the intesnity rates corresponding to the transient states has real and distinct characteristic roots.

Rust (1978)

13. Suppose $\{X(t), t \in IR^+\}$ is a finite Markov process on the state-space $\{0, 1, \ldots, M\}$ with state zero an absorbing state and the remaining states $1, \ldots, M$ transient. Assume that $i \leftrightarrow j$ and that $i \rightarrow 0$ for all $i, j = 1, \ldots, M$. Let

$$Q = \begin{pmatrix} 0 & 0_{1M} \\ C & B \end{pmatrix}$$

be the partitioned form of the intensity matrix, where B is $M \times M$ matrix, C is $M \times 1$ column vector and 0_{1M} is $1 \times M$ null row vector. Establish the following claims.

(i) The transition probability matrix of the process is

$$P(t) = \begin{pmatrix} 1 & 0_{1M} \\ C(t) & B(t) \end{pmatrix},$$

where $B(t) = \exp(Bt)$, $C(t) = E_{M1} - B(t)E_{M1}$.

(ii) The matrix B has a simple characteristic root $\lambda_1 < 0$ such that if λ is any other characteristic root of B, $Re(\lambda) < \lambda_1$. The corresponding right and left characteristic vectors $\xi = (u_1, \ldots, u_M)^T$ and $\eta = (v_1, \ldots, v_M)^T$ respectively of B have positive elements u_j and v_j, $j=1,\ldots,M$.

We may choose η such that $\eta^T E_{M1} = 1$ and also $\eta^T \xi = 1$.

Mandl (1960)

(iii) In view of results in (ii),

$$B(t) = \exp(t\lambda_1)\, \xi\eta^T + O(e^{t\lambda}) ,$$

where $\lambda < \lambda_1$.

(iv) Let $p_j(t) = \Pr[X(t) = j]$, $j = 0,1,\ldots, M$ and write

$$p(t) = [p_0(t), p_1(t), \ldots, p_M(t)]^T = [p_0(t), \alpha^T(t)]^T .$$

Define $d(t) = \alpha(t)/\{1 - p_0(t)\}$. The vector $d(t)$ is said to define a quasi-stationary distribution of the absorbing Markov process if $d(t) \equiv d$, $t \in IR^+$, with $d^T E_{M1} = 1$. We claim that $d = \eta$.

(v) For any initial distribution $(\pi_1, \ldots, \pi_M)$, $\pi_0 = 0$, $\pi_j \geq 0$, $\sum_{j=1}^{M} \pi_j = 1$, as $t \to \infty$,

$$p_j(t)/\{1 - p_0(t)\} \to v_j, \quad j = 1,\ldots,M.$$

If $\tau < t$,

$$\Pr[X(\tau) = j, \mid X(t) \neq 0] \to u_j v_j , \quad j = 1,\ldots,M.$$

(vi) For any initial distribution as in (v) and $\tau < t$,

$$\lim_{\tau\to\infty} \lim_{t\to\infty} \Pr[X(\tau) = j \mid X(t) \neq 0] \to u_j v_j .$$

(vii) Let $Z_{ij}(t)$ be the time spent in state j during $(0, t)$, given $X(0) = i$, $i, j = 1, \ldots, M$, and let Z_i denote the time to absorption in state 0, given $X(0) = i$. Then as $t \to \infty$

$$E\{Z_{ij}(t)/Z_i \mid Z_i > t\} \to u_j v_j , \quad j = 1, \ldots, M.$$

Darroch and Seneta (1967)

14. Darroch and Morris (1967) refer to the intensity matrix Q of a finite Markov process as a <u>log-stochastic matrix</u> since $\exp(Q)$ is a stochastic matrix. Show that if B is a square matrix with non-positive row sums, then $\exp(Bt)$ is sub-stochastic for all $t \in \mathbb{R}^+$. Such a matrix B may be called <u>log-sub-stochastic</u>. Let B be log-substochastic, then show that

(i) if $\exp(Bt) \to 0$ as $t \to \infty$, then B is non-singular

and

$$B^{-1} = - \int_0^\infty \exp(Bt)\, dt,$$

(ii) if $\exp(Bt) \not\to 0$, then B is singular.

Darroch and Morris (1967)

15. Let $A_i(t)$ denote the time spent in state i during $(0,t)$ by a finite Markov process with state-space $\{1, \ldots, M\}$. Let

$Y(t) = \sum_{i=1}^{m} d_i A_i(t)$, $m < M$, be a linear function of

$A_1(t), \ldots, A_m(t)$. Define

$$Z(a) = \begin{cases} \inf \{ t \mid Y(t) > a\} & \text{if } a < Y(\infty) \\ \infty \quad \text{if } Y(\infty) < \infty & \text{and } a \geq Y(\infty) . \end{cases}$$

Obtain the m.g.f. of $Z(a)$ and of $Z(a+b) - Z(a)$ if $b > 0$.

Darroch and Morris (1968)

16. Let $Y(t)$ and $Z(a)$ be defined as in Problem 15. Show that their m.g.f.'s satisfy the following identity :

$$\theta \int_0^\infty e^{-\theta a} E[e^{-\varphi Z(a)}]da = 1 - \varphi \int_0^\infty e^{-\varphi t} E[e^{-\theta Y(t)}]dt$$

valid when $\theta, \varphi > 0$.

Darroch (1966)

17. Let

$$Q^* = \begin{pmatrix} 0 & 0_{1M} & 0 \\ \alpha_0 & Q & \alpha_M \\ 0 & 0_{1M} & 0 \end{pmatrix}$$

be the intensity matrix of a finite Markov process on the state-space $\{0, 1, \ldots, M\}$ in which states 0 and M are absorbing and the transient states $1, \ldots, M-1$ communicate amongst themselves as well as lead to states 0 and M. Let τ_{jk} be the time spent in state k before absorption when the initial state is j and let $\tau_j = \sum_{k=1}^{M-1} \tau_{jk}$. In the notation of chapter 4 , let $K_{jk} = E[\tau_{jk}]$, $J_j = E(\tau_j)$,

$K = ((K_{jk}))$. Let K^*_{jk} denote the corresponding expected values for the $\mathbf{X}(R)$-chain. Show that

(i) $K_{jk} = K^*_{jk}/q_j$, $\quad j, k = 1, \ldots, M-1$,

(ii) $\Pr[\tau_{ij} = 0] = 1 - K_{ij} K^{-1}_{jj}$,

$$\Pr[0 < \tau_{ij} \leq t] = \{K_{ij}/K_{jj}\} \{1 - \exp(-tK^{-1}_{jj})\},$$

$$0 < t < \infty .$$

(iii) Let π_{ik} be the probability of absorption of the process in state K, $K = 0, M$ from the initial state i, $i = 1, \ldots, M-1$ and let $\boldsymbol{\pi}_k = (\pi_{1k}, \ldots, \pi_{M-1k})^T$, $k = 0, M$. Show that

$$\boldsymbol{\pi}_k = K\alpha_k , \quad k = 0, M .$$

(iv) Show that the Laplace transform of the joint distribution of $\tau_{i1}, \ldots, \tau_{iM-1}$ is

$$E\left[\prod_{j=1}^{M-1} \exp(-\theta_j \tau_{ij})\right]$$

$$= \eta_i^T (Q - \theta)^T Q E_{M-i,1} ,$$

where η_i is an $(M-1) \times 1$ vector with unity in the i-th position and zero everywhere else, and $\theta = \text{diag}\{\theta_1, \ldots, \theta_{M-1}\}$, $\theta_j > 0$, $j = 1, \ldots, M-1$.

(v) Obtain the Laplace transform of τ_i and show that its density is

$$f_i(t) = -\eta_i^T Q \exp(Qt) E_{M-1,1} , \quad t \in \mathbb{R}^+ ,$$

which is the density of the so-called matrix exponential distribution.

(vi) Obtain the mean and variance of τ_i.

Tavarè (1979)

18. Let $\{X(t), t \in IR^+\}$ be a finite Markov process with state-space $S = \{1, \ldots, M\}$ and intensity matrix Q which is assumed to have distinct characteristic roots $\lambda_1, \ldots, \lambda_M$. Let H be the M-square non-singular matrix whose column vectors are the right characteristic vectors of Q.

(i) Show that if

$$P(t_1, t_2) = ((\Pr[X(t_2) = j \mid X(t_1) = i])) , i, j = 1,\ldots,M,$$

$0 \leq t_1 < t_2 < \infty$, then

$$P(t_1, t_2) = \sum_{r=1}^{M} \exp\{\lambda_2(t_2 - t_1)\} H E_r H^{-1}$$

where E_r is a M-square matrix with 1 in the (r,r) position and zero everywhere else.

(ii) Let $p(t)$ be the $1 \times M$ row vector with $p_j(t) = \Pr[X(t) = j]$ in the j-th position, $j = 1, \ldots, M$. Show that

$$p(t) = \sum_{r=1}^{M} \exp(\lambda_r t)\, p(0)\, HE_r H^{-1}$$

(iii) Let the process be ergodic and $\lambda_1 = 0$. Then show that $\Pi = \lim_{t\to\infty} p(t) = H_1^{-1}$, where H_1^{-1} is the first row of H^{-1}.

(iv) Use the above results to obtain the covariance between $X(t_1)$ and $X(t_2)$, $t_1 < t_2$, when the initial distribution is (a) arbitrary and (b) the invariant distribution.

Reynolds (1972)

19. The Fix and Neyman model (cf. Example 4.8.2) of illness has been extended to the case of r diseases to study competing risks of illness. Suppose there are r diseases operating in a human population. An individual may be affected by any combination of the r diseases. Thus there are 2^r possible states of illness including the state of the healthy individual, not suffering from any disease. An individual may leave a state either through recovery or by being affected by new diseases. In a similar way there are 2^r possible states of death, according to the multiplicity of causes of death. Each of the 2^r states of death are absorbing states.

Suppose there are N individuals in state α at $t = 0$, $1 \leq \alpha \leq 2^r$. Let $N_{\alpha\beta}(t)$ denote the number of individuals in state β of illness at epoch t given that at $t = 0$, they were in state α. Similarly let $D_{\alpha\delta}(t)$ denote the deaths out of N_α individuals due to state δ of death.

The transitions among the 2^{r+1} states are governed by risks $s_{\alpha\beta}$ of illness and risks $v_{\alpha\delta}$ of death which are in fact intensity rates of transition between states α, β of illness and between state α of illness and δ of death.

Obtain expressions for the probability generating function of $N_{\alpha\beta}(t)$, $D_{\alpha\delta}(t)$ and the first two moments of length of stay in state β of illness during $(0, t)$ for an individual in state α at epoch 0.

Chiang (1961)

20. A general stochastic epidemic describes the infection of a group of N individuals called susceptibles into which a infectives are introduced at epoch $t = 0$. The state of the process at an epoch t can be described by the random vector $(X(t), Y(t))$, where $X(t)$ is the number of susceptibles and $Y(t)$ is the number of infectives at epoch t, $0 \leq X(t) + Y(t) \leq N + a$. Suppose $X(t) = x$, $Y(t) = y$, then during $(t, t+h)$, the transition intensities are specified by the following table

Transition during $(t, t+h)$	Probability
$(x, y) \to (x-1, y)$	$\beta xy\, h + o(h)$
$(x, y) \to (x, y-1)$	$\gamma yh + o(h)$
any other	$o(h)$.

In the above β is called the contact rate and γ is called the removal rate. Thus $(X(t), Y(t))$ is a two-dimensional finite Markov process on the state-space

$$\{(x,y) \ , \ x = 0, 1, \ldots, N+a, \ y = 0, 1, \ldots, N+a, \ x+y \leq N+a\}.$$

Let $p_{XY}(t) = \Pr[X(t) = x, Y(t) = y \mid X(0) = N, Y(0) = a]$ and define

$$f_x(z, t) = \sum_{y=0}^{N+a-x} z^y p_{xy}(t), \quad 0 \leq x \leq N.$$

Obtain partial differential equations satisfied by $f_x(z,t)$ and solve them by using direct recursive integration [cf. Siskind (1965)] , and by using Laplace transforms [cf. Gani (1965)].

21. In a reversible uni-molecular reaction $A \rightleftarrows B$, A molecules change into B molecules and B molecules change into A molecules. Suppose $X(t)$ is the number of A molecules at epoch t and $Y(t)$ is the number of B molecules with $X(t) + Y(t) = M$. A stochastic model for the chemical reaction assumes that $\{X(t), t \in \mathbb{R}^+\}$ is a birth-death process on $\{0, 1, \ldots, M\}$ with birth rate $\lambda_n = (M - n)\lambda$ and $\mu_n = n\mu$, $n = 0, 1, \ldots, M$, $\lambda > 0$, $\mu > 0$. Let $\Pr[X(0) = M] = 1$ and define

$$F(s, t) = \sum_{n=0}^{M} p_{Mn}(t)s^n.$$

Show that

$$\frac{\partial F(s,t)}{\partial t} = \{\mu + (\lambda-\mu)s + \lambda s^2\} \frac{\partial F(s,t)}{\partial s} + M\lambda(s-1) .$$

Solve this partial differential equation subject to the condition $F(s, 0) = s^M$. Hence or otherwise obtain the expected value $E(X(t))$ and the variance $V(X(t))$ of $X(t)$. Verify that for all $t > 0$

$$V[X(t)] = V(Y(t)) = E[X(t)]\, E[Y(t)]/ M ,$$

a relation predicted by statistical thermodynamics for the equilibrium state.

McQuarrie (1967), Tallis and Leslie (1969)

22. A finite birth-death process on the state-space $\{0, 1, \ldots, M\}$ with birth rates λ_i, $i = 0,1,\ldots, M-1$ and death rates μ_i, $i = 1,\ldots, M$ defined by

$$\lambda_i = \lambda(M - i) \{ i(1 - \alpha_1) + (M - i) \alpha_2\}/M$$

$$\mu_i = \mu\, i \{i\alpha_1 + (M - i) (1 - \alpha_2)\}/M$$

is used in Genetics to study changes in gene-frequencies in the presence of mutation for a randomly mating population of fixed size M. Obtain the transition probabilities and the invariant distribution for this process using spectral resolution of the intensity matrix.

Karlin and McGregor (1962, 1964)

23. Herbst (1963) considered a five state Markov process model to describe labour turn over in an industry. The states and the possible transitions are given in the following diagram.

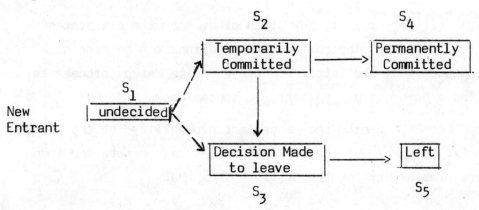

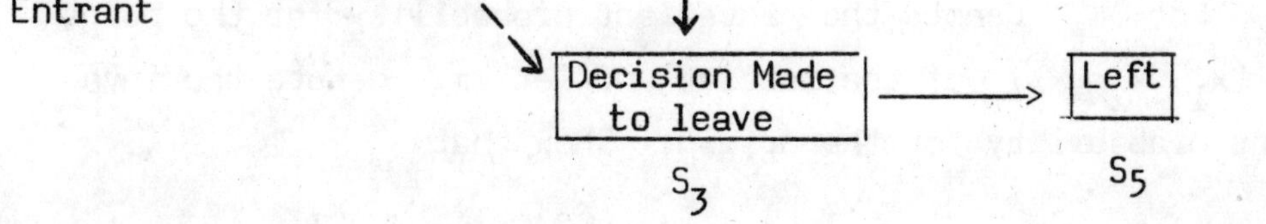

Thus the only non-zero intensity rates are r_{12}, r_{13}, r_{23}, r_{24}, r_{35}. Classify the states of this Markov process and obtain the proportion of those new-entrants who are likely to be permanently committed.

24. Suppose a particle moves on a finite three dimensional lattice which includes the origin. If the particle is at a point (x_1, x_2, x_3) of the lattice at an epoch t, then during (t, t+h], it can move to one of its six nearest neighbours with probabilities specified below.

State to which the particle moves from (x_1, x_2, x_3)	Probability
(x_1+1, x_2, x_3)	$\mu_E h + o(h)$
(x_1-1, x_2, x_3)	$\mu_W h + o(h)$
(x_1, x_2+1, x_3)	$\mu_N h + o(h)$
(x_1, x_2-1, x_3)	$\mu_S h + o(h)$
(x_1, x_2, x_3+1)	$\mu_U h + o(h)$
(x_1, x_2, x_3-1)	$\mu_D h + o(h)$
(x_1', x_2', x_3')	$o(h)$,

where (x_1', x_2', x_3') is any state other than the six states listed above it. Appropriate modifications can be made at the boundary of the lattice. Such a finite Markov process is called a homogeneous anisotropic random walk.

Let π_P denote the invariant probability for the point $P = (x_1, x_2, x_3)$ of the lattice and let π_0 denote the invariant probability for the origin. Show that

$$\pi_P = \{\mu_E/\mu_W\}^{x_1} \{\mu_N/\mu_S\}^{x_2} \{\mu_U/\mu_D\}^{x_3} \pi_0 .$$

Kramer (1959)

25. Suppose customers arrive at a service facility according to a Poisson process with rate λ. An arriving customer is a male with probability p and a female with probability $q = 1-p$, $0 < p < 1$, independently of the past of the process. If $X(t)$ is the sex of the last arrival before epoch t, show that $\{X(t),\ t \in \mathbb{R}^+\}$ is a two-state Markov process on the state space (M, F), with transition function

$$P(t) = \begin{pmatrix} p + q \exp(-\lambda t) & q - q \exp(-\lambda t) \\ p - p \exp(-\lambda t) & q + q \exp(-\lambda t) \end{pmatrix}.$$

(i) Obtain the likelihood function of p and λ based on a complete realization of the process over $[0, t]$.

(ii) Obtain the ML estimators of p and λ and discuss their asymptotic properties.

(iii) How will you test the hypotheses (a) $p = 1/2$ (b) $p = 1/2$, $\lambda \leq \lambda_0$ against their negations as the alternative hypotheses.

26. In the notation and terminology of section 5.5, suppose that the initial distribution of the irreducible finite Markov process is a continuous function in the topology of coordinatewise convergence of the transition probabilities and the transition intensities, and if for each $c \in (0, 1)$ the class of c-approximate ML estimators is uniformly consistent, then show that the identifiability condition holds for $\boldsymbol{\theta} = \boldsymbol{\theta}^0$.

Ranneby (1978)

27. Let $\{X(t),\ t \in \mathbb{R}^+\}$ be a birth-death process on the set of non-negative integers with birth rates $\lambda(i,\boldsymbol{\theta})$ and death rates $\mu(i,\boldsymbol{\theta})$ depending on a vector parameter $\boldsymbol{\theta}$ such that for all $\boldsymbol{\theta}$

$$\lambda(i,\boldsymbol{\theta}) > 0,\quad i \geq 0,\qquad \mu(0,\boldsymbol{\theta}) = 0,\qquad \mu(i,\boldsymbol{\theta}) > 0,\ i \geq 1,$$

$$\lambda(i,\boldsymbol{\theta})/\mu(i,\boldsymbol{\theta}) \leq c_1 < 1 \quad \text{for all} \quad i \geq k,$$

$$\lambda(i,\boldsymbol{\theta}) + \mu(i,\boldsymbol{\theta}) \geq c_2 > 0 \quad \text{for} \quad i \geq 0,$$

and $\lambda(i,\boldsymbol{\theta})/i \leq c_3 < \infty$,

where c_1, c_2, c_3 are constants.

(i) Show that such a birth-death process has an invariant distribution.

(ii) Obtain the likelihood $L_t(\boldsymbol{\theta})$ of $\boldsymbol{\theta}$ based on a complete observation of the process over $[0, t]$.

(iii) Show that the likelihood equations

$$\partial L_t(\theta)/\partial\theta_j = 0 , \quad j = 1, \ldots, r$$

have under certain regularity conditions, for sufficiently large t, a solution $\hat{\boldsymbol{\theta}}(t)$ with probability exceeding $1 - \varepsilon$, for any chosen $\varepsilon < 1$.

(iv) Show that the solution $\hat{\boldsymbol{\theta}}(t)$ is weakly consistent and that it provides a local maximum of the likelihood with probability exceeding $1 - \varepsilon$, provided t is sufficiently large.

(v) Show that under certain regularity conditions the asymptotic distribution of $\hat{\boldsymbol{\theta}}(t)$ is r-dimensional normal distribution.

Allen (1983)

28. Consider a birth-death process on $S = \{0, 1, \ldots, M\}$ with positive birth rates λ_j, $j = 0, 1, \ldots, M-1$ and positive death rates μ_j, $j = 1, \ldots, M$. Write down the likelihood function of λ_j and μ_j, $j \varepsilon S$, assuming that the initial distribution is known and the process is observed completely over $[0, t]$.

(i) Obtain the ML equation estimators of λ_j and μ_j, $j \varepsilon S$.

(ii) Obtain their joint asymptotic distribution.

(iii) Suggest procedures for testing the hypotheses

(a) $\lambda_j = \lambda_j^o$, $\mu_j = \mu_j^o$, $j \varepsilon S$,

(b) $\lambda_0 = \lambda_1 = \ldots = \lambda_{M-1}$, $\mu_0 = \mu_1 = \ldots = \mu_M$,

(c) $\lambda_0 = \lambda_1 = \ldots = \lambda_{M-1}$.

(iv) A variety of models are covered if we assume $\lambda_j = a(j)\lambda$, $\mu_j = b(j)\mu$, and $a(.)$ and $b(.)$ are known functions defined on S. Obtain the ML equation estimators of λ and μ and suggest test procedures for testing the composite hypothesis $\lambda = \mu$ against the alternative $\lambda \neq \mu$.

Wolff (1965)

29. An important method of observing a finite Markov process which yields limited information arises if the process is observed at a number of widely spaced time points. It is then assumed that we have a random sample from the invariant distribution of the process. Suppose such a sampling scheme

is adopted to observe a two-state Markov process. Obtain the ML estimators of the invariant probabilities $\lambda/(\lambda+\mu)$ and $\mu/(\lambda+\mu)$ and state their asymptotic distributions. Compare their efficiencies with those of the ML equation estimators based on a continuous large period of observation. Can one adopt this procedure of sampling over widely spaced observations to estimate λ and/or μ ?

Cox (1965)

30. A method frequently used in industry for sampling the performance of a group of machines is to observe some or all of them at objectively selected epochs which are not necessarily equidistant. This method is called snap-reading. If the operation of the machine is governed by a Markov process, write down the likelihood of the parameters of the process based on snap-reading. Obtain the ML equation estimators and study their asymptotic properties.

REFERENCES

Adke, S.R. and Manjunath, S.M. (1984). Sequential estimation for continuous time finite Markov processes, to appear in Communications in Statistics : Theory and Methods , Vol. 13, No. 9.

Aitchison, J. and Silvey, S.D. (1958). Maximum likelihood estimation of parameters subject to restraints, Ann.Math. Statist., 29, p. 813-828.

Albert, A.(1962). Estimating the infinitesimal generator of a continuous time finite state Markov process, Ann. Math. Statist., 38, p. 727-753.

Allen, O. B. (1983). Asymptotic properties of the maximum likelihood estimator for a class of birth-and-death processes admitting a unique stationary distribution, The Canadian J. Statist.,11, p. 109-118.

Apostol, T. M. (1975). Mathematical Analysis, Addison-Wesley.

Azema, J., Kaplan-Duflo, M. and Revuz, D. (1967). Méasure invariante sur les classes recurrentes des processus de Markov, Z. Wahrscheinlichkeitstheorie verw. Geb. 8, p. 157-181.

Bharucha-Reid, A. T. (1956). Note on the estimation of the number of states in a discrete Markov chain, Experientia, 12, p. 176-177.

Bhat, B. R. (1981). Modern Probability Theory, Wiley Eastern, New Delhi.

Bhat, U. N. (1972). Elements of Applied Stochastic Processes, John Wiley, New York.

Billingsley, P. (1961). Statistical Inference for Markov Processes, University of Chicago Press.

Billingsley, P. (1961a). Statistical methods in Markov chains, Ann. Math. Statist., 27, p. 1123-1129.

Birch, M. W. (1964). A new proof of the Pearson-Fisher theorem, Ann. Math. Statist., 35, p. 817-824.

Chiang, C. L. (1961). On the probability of death from specific causes in the presence of competing risks, Proc. 4th Berkely Symp. Math. Statist. Prob. Vol. IV, p. 169 -180 .

Chiang, C. L. (1968). Introduction to Stochastic Processes in Biostatistics , John Wiley, New York.

Chiang, C. L.and Raman, S. (1971). On a solution of Kolmogorov differential equations, Proceedings of the Fourth Conference on Probability Theory, Brasow, Romania, p. 129 -136.

Chung, K. L. (1967). Markov Chains with Stationary Transition Probabilities, Second Ed., Springer, Berlin.

Cinlar, E. (1975). Introduction to Stochastic Processes,Prentice Hall, New Jersey.

Cox, D. R. (1965). Some problems of statistical analysis connected with congestion, in Congestion Theory, Editors :W.L. Smith and W. E. Wilkinson, University of North Carolina Press.

Darroch, J. N. (1966). Identities for passage times with applications to recurrent events and homogeneous differential functions, J. Appl. Prob. 3, p. 435-444 .

Darroch, J. N. and Morris, K. W. (1967). Some passage - time generating functions for discrete-time and continuous - time finite Markov chains, J. Appl.Probability, 4, p.496-507.

Darroch, J. N. and Morris, K. W. (1968). Passage-time generating functions for continuous-time finite Markov chains, J. Appl. Prob., 5, p. 414-426.

Darroch, J. N. and Seneta, E. (1967). On quasi-stationary distributions in absorbing continuous time finite Markov chains, J. Appl. Prob., 4, p. 192-196.

Doob, J. L. (1953). Stochastic Processes, John Wiley, New York.

Feller, W. (1959). Non-Markovian process with the semi-group property, Ann. Math. Statist., 30, p. 1252-1253.

Feller, W. (1969). An Introduction to Probability Theory and Its Applications, Vol.II, Wiley Eastern, New Delhi.

Feller, W. (1972). An Introduction to Probability Theory and Its Applications, Vol. I, Wiley Eastern, New Delhi.

Fix, E. and Neyman, J. (1951). A simple stochastic model of recovery, relapse, death and loss of patients, Human Biology, 23, p. 205-241.

Gani, J. (1965). On a partial differential equation of epidemic theory, Biometrika, 52, p. 617-622.

Gnedenko, B. V. and Kovalenko, I. N. (1968). Introduction to Queueing Theory, Israel Programme for Scientific Translations, Jerusalem.

Grenander, U. (1981). Abstract Inference, John Wiley, New York.

Heathcote, C. R. and Moyal, J. E. (1959). The random walk in continuous time and its application to the theory of queues. Biometrika, 46, p. 400-411.

Herbst, P. G. (1963). Organisational commitment : a decision process model, Acta Sociologica, 7, p. 34-45.

Holgate, P. (1967). The size of elephant herds, Math. Gaz., 51, p. 302-304.

Iosifescu, M. (1980). Finite Markov Processes and Their Applications, John Wiley, New York.

Karlin, S. and McGregor, J. L. (1957). The differential equations of birth and death processes and the Stieltjes moment problem. Trans. Amer. Math. Soc. 85,p.489 - 546.

Karlin, S. and McGregor, J. L. (1962). On a genetic model of Moran, Proc. Cambridge Philos. Soc., 58, p. 299-311.

Karlin, S. and McGregor, J. L. (1964). On some stochastic models in genetics. In Stochastic Models in Medicine and Biology, Edited by Gurland,J., Univ.Wisconsin Press.

Karlin, S. and McGregor, J. L. (1965). Ehrenfest urn models, J. Appl. Prob., 2, p. 352-376.

Karlin, S. and Taylor, H. M. (1975). A First Course in Stochastic Processes, Second Edn. Academic Press, New York.

Kato, T. (1964). Perturbation Theory for Linear Operators, Springer-Verlag, New York.

Kemeny, J. G. and Snell, J. L. (1961). Finite continuous time Markov chains. Theor. Verojatnost i Primenen, 6,p.110-115.

Kemeny, J. G. and Snell, J. L. (1976). Finite Markov chains, Springer-Verlag, New York.

Kingman, J. F. C. (1962). The imbedding problem for finite Markov chains, Z. Wahrscheinlichkeitstheorie,1, p.14-24.

Kingman, J. F. C. (1963). The exponential decay of Markov transition probabilities, Proc. London Math. Soc. 13, p.337-358.

Kolmogorov, A. N. (1931). Sur le probleme d'attante, Recueil Mathematique Sbornik, 38, p. 101-106.

Kramer, H. P. (1959). Symmetrizable Markov matrices. Ann.Math. Statist., 30, 149-153.

Lamperti, J. (1977). Stochastic Processes, Springer - Verlag, New York.

Loeve, M. 1968). Probability Theory, 3rd Edn. East-West Student Edition.

Mandle, P. (1960). On the asymptotic behaviour of probabilities within groups of states of a homogeneous Markov process, Cas. pest. mat. 85, p. 448-456 (in Russian).

Manjunath, S. M. (1984). Optimal sequential estimation for ergodic birth-death processes, to appear in J. Roy. Statist. Soc. B , 46, part 3.

Markov, A. A. (1906). Extension of the law of large numbers to dependent events (in Russian), Bull. Soc. Phys.Math. Kazan (2), 15, p. 155-156.

McQuarrie, D. A. (1967). Stochastic approach to chemical kinetics,,J. Appl. Prob., 4, p. 413-478.

Medhi, J. (1982). Stochastic Processes, Wiley Eastern, New Delhi.

Moran, P.A.P. (1963). Some general results on random walks with genetic applications, J. Austral. Math. Soc., 3, p. 468-479.

Moran, P.A.P. (1968). An Introduction to Probability Theory, Clarendon Press, Oxford.

Palm, C. (1947). The distribution of repairmen in servicing automatic machines (in Swedish),Industritidningen Norden, 75, p. 75-80, 90-94, 119-123.

Parzen, E. (1962). Stochastic Processes, Holden-Day, San Francisco.

Pedler, P. J. (1971). Occupation times for a two state Markov chain, J. Appl. Prob., 8, 381-390.

Ranneby, B. (1978). On necessary and sufficient conditions for consistency of MLE's in Markov chain models, Scand. J. Statist., 5, p. 99-105.

Rao, C. R. (1973). Linear Statistical Inference and Its Applications, Second Ed., John Wiley, New York.

Reid, A. T. (1953). On stochastic processes in Biology, Biometrics, 9, 275-289.

Reynolds, J. F. (1972). Some theorems on the transient covariance of Markov chains. J. Appl. Prob., 9, p.214-218.

Rosenlund, S. I. (1978). Transition probabilities for truncated birth-death process, Scand. J. Statist.,5,p.119-122.

Royden, H. L. (1968). Real Analysis, Second Ed., Macmillan, New York.

Rust, P. F. (1978). The variance of duration of stay in an absorbing Markov process, J. Appl.Prob.,11,p. 572-577.

Siskind, V. (1965). A solution of the general stochastic epidemic, Biometrika, 52, p. 613-616.

Speakman, J. M. O. (1967). Two Markov chains with a common skeleton, Z. Wahrscheinlichkeitstheorie, 7, p. 224.

Stam, J. (1965). Derived stochastic processes, Campos. Math. Groningen, 17, p. 102-140.

Takacs, L. (1960) Stochastic Processes, Methuen's monographs, London.

Tallis, G. M. and Leslie, R. T. (1969). General models for r-molecular reactions, J. Appl. Prob. 6, p. 74-87.

Tavarè, S. (1979). A note on finite homogeneous continuous Markov chains. Biometrics, 35, p. 831-834.

Wasserman, S. (1980). Analysing social networks as stochastic processes, JASA, 75, p. 280-294.

Widder, D. V. (1946). The Laplace Transforms, Princeton Univ. Press, Princeton.

Wolff, R. W. (1965). Problems of statistical inference for birth and death queueing models. Operation Res., 13, p. 343-357.

Yeh, J. (1973). Stochastic Processes and the Wiener Integral, Marcel Dekker, New York.

Zacks, S. (1971). The Theory of Statistical Inference, John Wiley, New York.

AUTHOR INDEX

SUBJECT INDEX